Cambridge Tracts in Mathematics
and Mathematical Physics

General Editors
P. HALL, F.R.S., and F. SMITHIES, Ph.D.

No. 50

THE MATHEMATICS OF RADIATIVE TRANSFER

THE MATHEMATICS OF RADIATIVE TRANSFER

BY

I. W. BUSBRIDGE
Fellow of St Hugh's College, Oxford

CAMBRIDGE
AT THE UNIVERSITY PRESS
1960

CAMBRIDGE UNIVERSITY PRESS
Cambridge, New York, Melbourne, Madrid, Cape Town, Singapore, São Paulo, Delhi

Cambridge University Press
The Edinburgh Building, Cambridge CB2 8RU, UK

Published in the United States of America by Cambridge University Press, New York

www.cambridge.org
Information on this title: www.cambridge.org/9780521043984

First published 1960
This digitally printed version 2008

A catalogue record for this publication is available from the British Library

ISBN 978-0-521-04398-4 hardback
ISBN 978-0-521-09066-7 paperback

CONTENTS

Chapter 10. ANISOTROPIC SCATTERING

APPENDIX

NOTES ON FURTHER PROBLEMS

PREFACE

E. Hopf's tract on *Mathematical Problems of Radiative Equilibrium* (now out of print) has, for twenty-six years, provided the basis for almost all rigorous work on transfer theory. Meanwhile the subject has grown enormously, and only a few authors have dealt with questions concerning the existence and uniqueness of solutions. The theory of finite atmospheres, in particular, has been in a confused state for some years.

In writing this tract, I have incorporated (with modifications) those parts of Hopf's tract which have proved most valuable in the development of the subject and those parts which are needed in the subsequent analysis. I have tried throughout to keep the treatment as simple as possible and to obtain results which will be of practical use. By simplifying the treatment it has been possible to consider a larger number of topics, but even so there are many important aspects of the subject which have only been touched upon, and others which have been ignored.

I should like to express my sincere thanks to Professor E. C. Titchmarsh, who read and criticized a large part of the book. Although I have not adopted all his suggestions, the final text owes a great deal to him. I should also like to thank Dr J. B. Sykes, who obtained for me copies of de-classified papers on neutron diffusion and who has helped me with translations of Russian papers. Finally, my thanks are due to the printers for their careful and accurate work and to the staff of the Cambridge University Press for their helpfulness and courtesy.

I. W. B.

ST HUGH'S COLLEGE
OXFORD
January 1960

PART I

AUXILIARY MATHEMATICS

CHAPTER 1

THE EQUATION OF TRANSFER

1. Introduction

Most of the completely solved problems in transfer theory either are concerned with atmospheres (or media) stratified in plane parallel layers, or they can be reduced to equations occurring in the theory of plane parallel atmospheres, or they can be solved by methods used for such atmospheres, and we shall therefore confine our attention to the plane case. The radiation field of a stellar atmosphere is always assumed to be stationary (i.e. independent of the time). A note on time-dependent problems is given in § 53.

Every problem is essentially that of solving the 'equation of transfer' (an integro-differential equation) subject to given boundary conditions. By certain standard techniques, the problem can be reduced to the solution of an integral equation. This reduction is given in §§ 4–7. In §§ 2–3 we give a brief derivation of the equation of transfer for a plane parallel atmosphere. This is based on 'intensity of radiation', a term which is introduced without a precise definition. For a more fundamental treatment, reference should be made to one of the standard works on radiative transfer.†

2. Notation

The energy flow at a point P in the radiation field of a plane parallel atmosphere is specified by the *intensity*. This is a function of the geometrical depth x of P below the upper surface of the atmosphere, of the direction of flow considered, and of the frequency ν of the radiation. In traversing a length

† See, for example, Chandrasekhar [1], chap. I.

ds at P where the density is ρ, the intensity I_ν of the radiation will be weakened by absorption of an amount $\rho\kappa_\nu I_\nu ds$, where κ_ν is the *mass absorption coefficient*. It will gain by emission in the element considered and by scattering of radiation from other directions into that of ds. Variations in frequency will not be considered and the suffix ν will therefore be omitted.

In transfer problems it is usual to introduce a depth variable τ, called the *optical depth*. It is defined by

$$\frac{d\tau}{dx} = \kappa\rho, \tag{2·1}$$

with $\tau = 0$ in the upper surface of the atmosphere.

Let μ denote the cosine of the angle θ which a given direction $\mathbf{l}$ makes with the outwards-drawn normal $\mathbf{n}$ to the surface $\tau = 0$, and let ϕ denote the azimuth of this direction referred to a fixed plane through $\mathbf{n}$. At a depth τ below the surface, the intensity of the radiation (of frequency ν) in the direction $\mathbf{l}$ will be denoted by $I(\tau, \mu, \phi)$. The notation $I(\tau, +\mu, \phi)$ and $I(\tau, -\mu, \phi)$ $(0 \leqslant \mu \leqslant 1)$ will be used when it is desirable to emphasize that $\mathbf{l}$ makes an acute angle $\cos^{-1}\mu$ with $\mathbf{n}$ or with $-\mathbf{n}$. In axially symmetric problems the variable ϕ will be omitted.

It will be assumed that radiation is scattered (without change of frequency) from the direction (μ, ϕ) into the direction (μ', ϕ') according to a *phase function* depending upon the cosine of the angle Θ between the two directions, i.e. upon

$$\cos\Theta = \mu\mu' + (1-\mu^2)^{\frac{1}{2}}(1-\mu'^2)^{\frac{1}{2}}\cos(\phi-\phi'), \tag{2·2}$$

and possibly upon the optical depth τ. The phase function will be denoted by $p(\mu, \phi; \mu', \phi'; \tau)$. It is essentially positive and must satisfy the relations

$$\left.\begin{aligned} p(\mu, \phi; \mu', \phi'; \tau) &= p(\mu', \phi'; \mu, \phi; \tau), \\ p(-\mu, \phi; \mu', \phi'; \tau) &= p(\mu, \phi; -\mu', \phi'; \tau). \end{aligned}\right\} \tag{2·3}$$

If $I(\tau, \mu, \phi)$ is the incident intensity at one end of an element of length ds in the direction (μ, ϕ), then

$$(\kappa\rho\, ds/4\pi)\, I(\tau, \mu, \phi)\, p(\mu, \phi; \mu', \phi'; \tau) \tag{2·4}$$

is the intensity in the direction (μ', ϕ') due to scattering in the element. On integrating over all directions (μ', ϕ'), we obtain the total reduction in the intensity in the (μ, ϕ) direction due to scattering. If scattering alone is present, the reduction must be $\kappa\rho I(\tau, \mu, \phi)\, ds$ (by definition of κ) and energy is conserved, but if some radiation is 'truly absorbed' (i.e. absorbed and re-emitted at another frequency) the total reduction due to scattering will be less than $\kappa\rho I(\tau, \mu, \phi)\, ds$. Hence

$$\frac{1}{4\pi}\int_{-1}^{1} d\mu' \int_{0}^{2\pi} p(\mu, \phi;\ \mu', \phi';\ \tau)\, d\phi' = \omega_0, \qquad (2\cdot5)$$

where $0 < \omega_0 \leqslant 1$. Since energy is conserved when $\omega_0 = 1$, this is known as the *conservative case*.

Except in § 51, it will be assumed that the phase function is independent of τ and that it can be expanded in the form

$$p(\mu, \phi;\ \mu', \phi') = \sum_{n=0}^{N} \omega_n P_n(\cos\Theta), \qquad (2\cdot6)$$

where $P_n(x)$ is the Legendre polynomial of order n, and N is a fixed integer. On using the addition theorem for Legendre polynomials, this becomes

$$p(\mu, \phi;\ \mu', \phi') = \sum_{n=0}^{N} \omega_n \left\{ P_n(\mu)\, P_n(\mu') + 2\sum_{m=1}^{n} c_{mn}\, P_n^m(\mu)\, P_n^m(\mu') \cos m(\phi - \phi') \right\}, \qquad (2\cdot7)$$

where $$c_{mn} = (n-m)!/(n+m)!. \qquad (2\cdot8)$$

Then relation (2·5) is true, ω_0 being a constant, viz. the first term of the expansion (2·6). It is known as the 'albedo for single scattering'. From (2·7) we also have the important relation

$$\frac{1}{4\pi}\int_{-1}^{1} \mu'\, d\mu' \int_{0}^{2\pi} p(\mu, \phi;\ \mu', \phi')\, d\phi' = \tfrac{1}{3}\omega_1 \mu. \qquad (2\cdot9)$$

When $N = 0$, then $p(\mu, \phi;\ \mu', \phi') = \omega_0$ and scattering is said to be *isotropic*. It may happen that scattering is isotropic at any given depth, i.e. that

$$p(\mu, \phi; \mu', \phi';\ \tau) = \omega_0(\tau), \qquad (2\cdot10)$$

where $0 < \omega_0(\tau) \leqslant 1$. Then much of the following analysis continues to hold. This case is considered briefly in § 51.

3. The equation of transfer

Consider the difference in the intensity of the radiation at the ends of a small cylinder of length ds in the direction (μ, ϕ). On taking account of signs, and using (2·1), we have

$$ds = -dx/\mu = -d\tau/\mu\kappa\rho. \tag{3·1}$$

The change in the intensity will be due to (i) absorption of an amount $\kappa\rho I(\tau, \mu, \phi)\, ds$, (ii) scattering from other directions into the given one of an amount (by (2·4) and (2·3))

$$\frac{\kappa\rho\, ds}{4\pi} \int_{-1}^{1} d\mu' \int_{0}^{2\pi} I(\tau, \mu', \phi')\, p(\mu, \phi;\, \mu', \phi')\, d\phi',$$

(iii) emission in the element of an amount $\kappa\rho B_1(\tau)\, ds$, where $B_1(\tau)$ is a known function which is independent of direction. Hence

$$\frac{\partial}{\partial s} I(\tau, \mu, \phi) = -\kappa\rho I(\tau, \mu, \phi) + \kappa\rho \mathfrak{J}(\tau, \mu, \phi), \tag{3·2}$$

where the *source function* $\mathfrak{J}(\tau, \mu, \phi)$ is given by

$$\mathfrak{J}(\tau, \mu, \phi) = \frac{1}{4\pi} \int_{-1}^{1} d\mu' \int_{0}^{2\pi} I(\tau, \mu', \phi')\, p(\mu, \phi;\, \mu', \phi')\, d\phi' + B_1(\tau). \tag{3·3}$$

On substituting for ds from (3·1) into (3·2), we obtain the equation of transfer

$$\mu \frac{\partial}{\partial \tau} I(\tau, \mu, \phi) = I(\tau, \mu, \phi) - \mathfrak{J}(\tau, \mu, \phi). \tag{3·4}$$

Equations (3·3) and (3·4) give an integro-differential equation for $I(\tau, \mu, \phi)$. This has to be solved subject to appropriate boundary conditions. If the atmosphere is of finite optical depth τ_1, $I(\tau, \mu, \phi)$ will usually be known when $\tau = 0$ and $\tau = \tau_1$. If it is of infinite depth, the condition for $\tau = \tau_1$ will be replaced by one limiting the rate of increase of $I(\tau, \mu, \phi)$, or of $\mathfrak{J}(\tau, \mu, \phi)$, for large τ. Alternatively, the net flux (see § 6) through the bounding surfaces may be given.

By their physical definitions, $I(\tau, \mu, \phi)$ and $B_1(\tau)$ are non-negative functions. It follows from (3·3) and the positivity of the phase function that $\mathfrak{J}(\tau, \mu, \phi)$ is also non-negative. In general, all these functions will be continuous for $\tau > 0$.

4. Finite atmospheres

Let radiation fall on the surfaces $\tau = 0$ and $\tau = \tau_1$ of intensities $I_0(\mu, \phi)$ and $I_1(\mu, \phi)$ such that, for $0 \leqslant \mu \leqslant 1$,

$$\left.\begin{aligned} I(0, -\mu, \phi) &= I_0(\mu, \phi), \\ I(\tau_1, +\mu, \phi) &= I_1(\mu, \phi). \end{aligned}\right\} \tag{4·1}$$

The solution of (3·4), subject to these boundary conditions and treating $\mathfrak{J}(\tau, \mu, \phi)$ as known, is

$$\begin{aligned} I(\tau, +\mu, \phi) = {} & I_1(\mu, \phi) \exp[-(\tau_1 - \tau)/\mu] \\ & + \int_\tau^{\tau_1} \mathfrak{J}(t, \mu, \phi) \exp[-(t-\tau)/\mu] \frac{dt}{\mu}, \end{aligned} \tag{4·2}$$

$$\begin{aligned} I(\tau, -\mu, \phi) = {} & I_0(\mu, \phi) \exp(-\tau/\mu) \\ & + \int_0^\tau \mathfrak{J}(t, -\mu, \phi) \exp[-(\tau - t)/\mu] \frac{dt}{\mu}, \end{aligned} \tag{4·3}$$

where $0 \leqslant \mu \leqslant 1$ in each equation.

4·1. In the case of finite atmospheres, we shall only consider axially symmetric problems with isotropic scattering. Then the phase function is ω_0, and ϕ is absent from the above equations. By (3·3) the source function, which is independent of μ, is given by

$$\mathfrak{J}(\tau) = \tfrac{1}{2}\omega_0 \int_{-1}^{1} I(\tau, \mu')\, d\mu' + B_1(\tau). \tag{4·4}$$

On writing this in the form

$$\mathfrak{J}(\tau) = \tfrac{1}{2}\omega_0 \int_0^1 I(\tau, +\mu')\, d\mu' + \tfrac{1}{2}\omega_0 \int_0^1 I(\tau, -\mu')\, d\mu' + B_1(\tau), \tag{4·5}$$

substituting from (4·2) and (4·3) and inverting the order of the t and μ' integrals, we get

$$\mathfrak{J}(\tau) = \tfrac{1}{2}\omega_0 \int_0^{\tau_1} \mathfrak{J}(t)\, E_1(|t - \tau|)\, dt + B(\tau), \tag{4·6}$$

where $$E_n(\tau) = \int_1^\infty x^{-n} \exp(-\tau x)\, dx \quad (n = 1, 2, \ldots) \tag{4·7}$$

and $B(\tau)$ is a known function given by

$$B(\tau) = B_1(\tau) + \tfrac{1}{2}\omega_0 \int_0^1 I_1(\mu') \exp[-(\tau_1 - \tau)/\mu']\, d\mu'$$

$$+ \tfrac{1}{2}\omega_0 \int_0^1 I_0(\mu') \exp(-\tau/\mu')\, d\mu'. \tag{4·8}$$

The integral operator $\bar{\Lambda}$ will be defined by

$$\bar{\Lambda}_\tau \{f(t)\} = \frac{1}{2} \int_0^{\tau_1} f(t)\, E_1(|t - \tau|)\, dt. \tag{4·9}$$

In order to avoid confusion with the intensity I, the identity operator will be denoted by 1. Then (4·6) can be written

$$(1 - \omega_0 \bar{\Lambda})_\tau \{\mathfrak{J}(t)\} = B(\tau). \tag{4·10}$$

We shall call this the *Milne equation* of the problem. When it has been solved for the source function $\mathfrak{J}(\tau)$, $I(\tau, \mu)$ can be found from (4·2) and (4·3). In general, the emergent intensities $I(0, +\mu)$ and $I(\tau_1, -\mu)$ are required. These are given by

$$I(0, +\mu) = I_1(\mu) \exp(-\tau_1/\mu) + \int_0^{\tau_1} \mathfrak{J}(t) \exp(-t/\mu) \frac{dt}{\mu}, \tag{4·11}$$

and

$$I(\tau_1, -\mu) = I_0(\mu) \exp(-\tau_1/\mu) + \int_0^{\tau_1} \mathfrak{J}(t) \exp[-(\tau_1 - t)/\mu] \frac{dt}{\mu}. \tag{4·12}$$

5. Semi-infinite atmospheres

An atmosphere is called semi-infinite if it exists for all $\tau \geqslant 0$. Let $\tau' > \tau \geqslant 0$ and $0 \leqslant \mu \leqslant 1$; then corresponding to (4·2), we have

$$I(\tau, +\mu, \phi) \exp(-\tau/\mu) - I(\tau', +\mu, \phi) \exp(-\tau'/\mu)$$

$$= \int_\tau^{\tau'} \mathfrak{J}(t, \mu, \phi) \exp(-t/\mu) \frac{dt}{\mu}. \tag{5·1}$$

Since $\mathfrak{J}(t, \mu, \phi) \geqslant 0$, the right-hand side of (5·1) is an increasing function of τ' (in the wide sense). Hence $I(\tau', +\mu, \phi) \exp(-\tau'/\mu)$ is a non-negative, decreasing function of τ', and it therefore tends to a non-negative limit $i(\mu, \phi)$ as $\tau' \to \infty$. Thus the integral on the right of (5·1) converges as $\tau' \to \infty$. Also

$$I(\tau, +\mu, \phi) \geqslant i(\mu, \phi) \exp(\tau/\mu) \geqslant i(\mu, \phi) \exp \tau, \qquad (5\cdot2)$$

since $0 \leqslant \mu \leqslant 1$.

From (5·2), (3·3) and the positivity of the functions, it follows that, when $0 \leqslant \mu \leqslant 1$,

$$\mathfrak{J}(\tau, \mu, \phi) \geqslant \frac{1}{4\pi} \exp \tau \int_0^1 d\mu' \int_0^{2\pi} i(\mu', \phi') p(\mu, \phi; \mu', \phi') \, d\phi'.$$

Hence if $\mathfrak{J}(\tau, \mu, \phi)$ is of smaller order than $\exp \tau$ for large τ and $0 \leqslant \mu \leqslant 1$, then $i(\mu, \phi)$ must be zero except, possibly, in a set of measure zero. If such a set exists, $I(\tau, +\mu, \phi)$ can be adjusted in that set to make $i(\mu, \phi)$ zero everywhere.

We shall assume that, as $\tau \to \infty$,

$$\mathfrak{J}(\tau, \mu, \phi) = O(\exp(a\tau)) \qquad (5\cdot3)$$

uniformly for $-1 \leqslant \mu \leqslant 1$, $0 \leqslant \phi \leqslant 2\pi$, where $0 \leqslant a < 1$. Then $i(\mu, \phi) = 0$ and (5·1) gives, on letting $\tau' \to \infty$,

$$I(\tau, +\mu, \phi) = \int_\tau^\infty \mathfrak{J}(t, \mu, \phi) \exp[-(t-\tau)/\mu] \frac{dt}{\mu}. \qquad (5\cdot4)$$

This replaces equation (4·2) when the atmosphere is semi-infinite. Equation (4·3) is unchanged if the boundary condition at $\tau = 0$ is

$$I(0, -\mu, \phi) = I_0(\mu, \phi) \quad (0 \leqslant \mu \leqslant 1). \qquad (5\cdot5)$$

Equation (3·3) can be written

$$\mathfrak{J}(\tau, \mu, \phi) = \frac{1}{4\pi} \int_0^1 d\mu' \int_0^{2\pi} p(\mu, \phi; \mu', \phi') I(\tau, +\mu', \phi') \, d\phi'$$
$$+ \frac{1}{4\pi} \int_0^1 d\mu' \int_0^{2\pi} p(\mu, \phi; -\mu', \phi') \times I(\tau, -\mu', \phi') \, d\phi' + B_1(\tau). \qquad (5\cdot6)$$

On substituting from (5·4) and (4·3) and inverting the orders of integration,† we get the following equation for $\mathfrak{J}(\tau, \mu, \phi)$:

† This is justified if $\mathfrak{J}(\tau, \mu, \phi)$ satisfies (5·3).

$$\mathfrak{J}(\tau,\mu,\phi) = \frac{1}{4\pi}\int_0^1 \frac{d\mu'}{\mu'}\int_0^{2\pi} p(\mu,\phi;\,\mu',\phi')\,d\phi'$$
$$\times\int_\tau^\infty \mathfrak{J}(t,\mu',\phi')\exp[-(t-\tau)/\mu']\,dt$$
$$+\frac{1}{4\pi}\int_0^1 \frac{d\mu'}{\mu'}\int_0^{2\pi} p(\mu,\phi;\,-\mu',\phi')\,d\phi'$$
$$\times\int_0^\tau \mathfrak{J}(t,\,-\mu',\phi')\exp[-(\tau-t)/\mu']\,dt + B(\tau,\mu,\phi), \tag{5·7}$$

where $B(\tau,\mu,\phi)$ is a known function given by

$$B(\tau,\mu,\phi) = B_1(\tau)+\frac{1}{4\pi}\int_0^1 d\mu'\int_0^{2\pi} p(\mu,\phi;\,-\mu',\phi')\,I_0(\mu',\phi')$$
$$\times\exp(-\tau/\mu')\,d\phi'. \tag{5·8}$$

Equation (5·7) is the Milne equation of the problem. It has to be solved for the source function $\mathfrak{J}(\tau,\mu,\phi)$.

In (5·7) and (5·8) all the functions are non-negative and it therefore follows that

$$B_1(\tau)\leqslant B(\tau,\mu,\phi)\leqslant \mathfrak{J}(\tau,\mu,\phi). \tag{5·9}$$

Hence, by (5·3), we must have

$$B_1(\tau) = O(\exp(a\tau)) \quad \text{as} \quad \tau\to\infty. \tag{5·10}$$

5·1. When there is isotropic scattering and axial symmetry, the source function is given by (4·4). Then (5·7) reduces to the infinite analogue of (4·6), viz.

$$\mathfrak{J}(\tau) = \tfrac{1}{2}\omega_0\int_0^\infty \mathfrak{J}(t)\,E_1(|t-\tau|)\,dt + B(\tau), \tag{5·11}$$

where $$B(\tau) = B_1(\tau)+\tfrac{1}{2}\omega_0\int_0^1 I_0(\mu')\exp(-\tau/\mu')\,d\mu'. \tag{5·12}$$

When $B(\tau)=0$, (5·11) is known as *Milne's First Integral Equation*.†

Equation (5·11) can be written in the form

$$(1-\omega_0\Lambda)_\tau\{\mathfrak{J}(t)\} = B(\tau), \tag{5·13}$$

† See Milne [1], equation (65), when $\omega_0 = 1$. It was first derived in Milne [2].

where Λ is Hopf's operator (see Hopf [1]) defined by

$$\Lambda_\tau\{f(t)\} = \frac{1}{2}\int_0^\infty f(t)\,E_1(|t-\tau|)\,dt. \tag{5·14}$$

From (5·4), the emergent intensity is

$$I(0, +\mu) = \int_0^\infty \mathfrak{J}(t)\exp(-t/\mu)\frac{dt}{\mu} = \mathfrak{L}_{1/\mu}\{\mathfrak{J}(t)\}, \tag{5·15}$$

where $\mathfrak{L}$ is the Laplace operator defined by†

$$\mathfrak{L}_s\{f(t)\} = s\int_0^\infty f(t)\exp(-st)\,dt. \tag{5·16}$$

6. The net flux

The rate of flow of energy (of frequency ν) across unit area of the plane at depth τ below the surface is measured by

$$\pi F(\tau) = \int_{-1}^{1}\mu\,d\mu\int_0^{2\pi} I(\tau,\mu,\phi)\,d\phi. \tag{6·1}$$

When there is axial symmetry, this becomes

$$\pi F(\tau) = 2\pi\int_{-1}^{1} I(\tau,\mu)\,\mu\,d\mu. \tag{6·2}$$

$F(\tau)$ is called the *net flux at depth* τ.

Let $\bar{I}(\tau)$ denote the mean value of the intensity over all directions (μ,ϕ), so that

$$\bar{I}(\tau) = \frac{1}{4\pi}\int_{-1}^{1} d\mu\int_0^{2\pi} I(\tau,\mu,\phi)\,d\phi. \tag{6·3}$$

Then on integrating (3·4) with respect to μ and ϕ, we get

$$\pi(d/d\tau)\,F(\tau) = 4\pi\bar{I}(\tau) - \int_{-1}^{1} d\mu\int_0^{2\pi}\mathfrak{J}(\tau,\mu,\phi)\,d\phi. \tag{6·4}$$

But from (3·3), on using (2·3) and (2·5),

$$\begin{aligned}\int_{-1}^{1} d\mu\int_0^{2\pi}\mathfrak{J}(\tau,\mu,\phi)\,d\phi &= \omega_0\int_{-1}^{1} d\mu'\int_0^{2\pi} I(\tau,\mu',\phi')\,d\phi' + 4\pi B_1(\tau)\\ &= 4\pi\omega_0\bar{I}(\tau) + 4\pi B_1(\tau).\end{aligned} \tag{6·5}$$

† The introduction of the factor s before the integral is usual in transfer theory because of the relation (5·15).

Hence (6·4) becomes

$$(d/d\tau)\,F(\tau) = 4(1-\omega_0)\,\bar{I}(\tau) - 4B_1(\tau). \tag{6·6}$$

When $\omega_0 = 1$ and $B_1(\tau) = 0$, (6·6) has the solution

$$F(\tau) = \text{constant} = F. \tag{6·7}$$

This is known as the *flux integral in the conservative case.*

6·1. *Expressions for $F(\tau)$ when $\omega_0 = 1$.* We now suppose that the atmosphere is semi-infinite. (Similar results are easily found for a finite atmosphere.) When the incident radiation $I_0(\mu, \phi)$ is zero, on substituting from (4·3) and (5·4) into (6·1), we get

$$\pi F(\tau) = \int_0^1 d\mu \int_0^{2\pi} d\phi \int_\tau^\infty \mathfrak{J}(t, \mu, \phi) \exp\left[-(t-\tau)/\mu\right] dt$$
$$- \int_0^1 d\mu \int_0^{2\pi} d\phi \int_0^\tau \mathfrak{J}(t, -\mu, \phi) \exp\left[-(\tau-t)/\mu\right] dt. \tag{6·8}$$

In the axially symmetric case in which $I_0(\mu) = 0$ and $B_1(\tau) = 0$, (6·7) and (6·8) give

$$F = 2\int_0^1 d\mu \int_\tau^\infty \mathfrak{J}(t, \mu) \exp\left[-(t-\tau)/\mu\right] dt$$
$$- 2\int_0^1 d\mu \int_0^\tau \mathfrak{J}(t, -\mu) \exp\left[-(\tau-t)/\mu\right] dt, \tag{6·9}$$

and in particular (taking $\tau = 0$)

$$F = 2\int_0^1 d\mu \int_0^\infty \mathfrak{J}(t, \mu) \exp(-t/\mu)\, dt. \tag{6·10}$$

The integral equation (6·9) can also be obtained by integrating the corresponding Milne equation (cf. (5·7))

$$\mathfrak{J}(\tau, \mu) = \frac{1}{2}\int_0^1 p(\mu, \mu') \frac{d\mu'}{\mu'} \int_\tau^\infty \mathfrak{J}(t, \mu') \exp\left[-(t-\tau)/\mu'\right] dt$$
$$+ \frac{1}{2}\int_0^1 p(\mu, -\mu') \frac{d\mu'}{\mu'} \int_0^\tau \mathfrak{J}(t, -\mu') \exp\left[-(\tau-t)/\mu'\right] dt \tag{6·11}$$

with respect to μ and τ over $(-1, 1)$ and $(0, \tau)$. Thus a solution of (6·11), for which the flux is F, will also be a solution of (6·9).

When the scattering is isotropic, (6·9) reduces to†

$$F = 2\int_\tau^\infty \mathfrak{J}(t)\,E_2(t-\tau)\,dt - 2\int_0^\tau \mathfrak{J}(t)\,E_2(\tau-t)\,dt, \qquad (6\cdot12)$$

since the source function is then independent of μ. The corresponding Milne equation is [$\omega_0 = 1$, $B(\tau) = 0$ in (5·13)]

$$\mathfrak{J}(\tau) = \Lambda_\tau\{\mathfrak{J}(t)\}, \qquad (6\cdot13)$$

and (6·12) can be obtained by integrating this over $(0,\tau)$.

7. The K-integral

Let
$$K(\tau) = \frac{1}{4\pi}\int_{-1}^{1}\mu^2\,d\mu\int_0^{2\pi} I(\tau,\mu,\phi)\,d\phi. \qquad (7\cdot1)$$

On multiplying (3·4) by μ and integrating with respect to ϕ and μ, we get

$$4\pi(d/d\tau)\,K(\tau) = \int_{-1}^{1}\mu\,d\mu\int_0^{2\pi} I(\tau,\mu,\phi)\,d\phi - \int_{-1}^{1}\mu\,d\mu\int_0^{2\pi}\mathfrak{J}(\tau,\mu,\phi)\,d\phi. \qquad (7\cdot2)$$

The first integral on the right is $\pi F(\tau)$. From (3·3), (2·3) and (2·9) it follows that

$$\int_{-1}^{1}\mu\,d\mu\int_0^{2\pi}\mathfrak{J}(\tau,\mu,\phi)\,d\phi = \tfrac{1}{3}\omega_1\int_{-1}^{1}\mu'\,d\mu'\int_0^{2\pi} I(\tau,\mu',\phi')\,d\phi' = \tfrac{1}{3}\omega_1\,\pi F(\tau). \qquad (7\cdot3)$$

Hence (7·2) becomes

$$4(d/d\tau)\,K(\tau) = (1-\tfrac{1}{3}\omega_1)\,F(\tau). \qquad (7\cdot4)$$

When $\omega_0 = 1$ and $B_1(\tau) = 0$, $F(\tau)$ is constant and therefore

$$K(\tau) = \tfrac{1}{4}(1-\tfrac{1}{3}\omega_1)\,F\tau + K(0). \qquad (7\cdot5)$$

This is known as the *K-integral in the conservative case.*‡ In particular, when scattering is isotropic, $\omega_1 = 0$ and

$$K(\tau) = \tfrac{1}{4}F\tau + K(0). \qquad (7\cdot6)$$

† Milne [1], equation (162).

‡ This was first given (for isotropic scattering) by Eddington in [1], equation (225·7).

7·1. *Expressions for $K(\tau)$ when $\omega_0 = 1$.* We again assume that the atmosphere is semi-infinite. When $I_0(\mu, \phi) = 0$, (7·1), (4·3) and (5·4) give the relation

$$K(\tau) = \frac{1}{4\pi}\int_0^1 \mu\, d\mu \int_0^{2\pi} d\phi \int_\tau^\infty \mathfrak{J}(t, \mu, \phi) \exp\left[-(t-\tau)/\mu\right] dt$$
$$+\frac{1}{4\pi}\int_0^1 \mu\, d\mu \int_0^{2\pi} d\phi \int_0^\tau \mathfrak{J}(t, -\mu, \phi) \exp\left[-(\tau-t)/\mu\right] dt. \tag{7·7}$$

When there is axial symmetry and $B_1(\tau) = 0$, $I_0(\mu) = 0$, (7·5) and (7·7) give the following integral equation for $\mathfrak{J}(\tau, \mu)$:

$$\frac{1}{2}\int_0^1 \mu\, d\mu \int_\tau^\infty \mathfrak{J}(t, \mu) \exp\left[-(t-\tau)/\mu\right] dt$$
$$+\frac{1}{2}\int_0^1 \mu\, d\mu \int_0^\tau \mathfrak{J}(t, -\mu) \exp\left[-(\tau-t)/\mu\right] dt$$
$$= \tfrac{1}{4}(1-\tfrac{1}{3}\omega_1) F\tau + K(0), \tag{7·8}$$

where
$$K(0) = \frac{1}{2}\int_0^1 \mu\, d\mu \int_0^\infty \mathfrak{J}(t, \mu) \exp(-t/\mu)\, dt. \tag{7·9}$$

If the scattering is isotropic, (7·8) reduces to

$$\frac{1}{2}\int_0^\infty \mathfrak{J}(t)\, E_3(|t-\tau|)\, dt = \tfrac{1}{4}F\tau + K(0). \tag{7·10}$$

CHAPTER 2

THE H-FUNCTIONS

8. Introduction

The solutions of most transfer problems in semi-infinite atmospheres can be reduced to expressions involving the H-functions of S. Chandrasekhar.† These satisfy integral equations of the form

$$\frac{1}{H(\mu)} = 1-\mu\int_0^1 \frac{\Psi(x)\,H(x)}{\mu+x}\,dx, \tag{8·1}$$

where $\Psi(x)$ is a known function which is non-negative for $0 \leqslant x \leqslant 1$ and satisfies the condition‡

$$\psi_0 \equiv \int_0^1 \Psi(x)\,dx \leqslant \tfrac{1}{2}. \tag{8·2}$$

$\Psi(x)$ is known as the *characteristic function*. The case of equality in (8·2) will be called the *conservative case*. Later this will be found to correspond to the conservative case defined in § 2.

It is easy to show from (8·1) (see § 10) that, if μ is complex and does not lie in the interval $(-1, 1)$,

$$1/\{H(\mu)\,H(-\mu)\} = T(\mu), \tag{8·3}$$

where

$$T(\mu) = 1-2\mu^2\int_0^1 \frac{\Psi(x)}{\mu^2-x^2}\,dx. \tag{8·4}$$

It follows that the singularities of $H(\mu)$ in the complex plane are related to the singularities and zeros of $T(\mu)$. These are studied in § 9.

The method by which the existence of $H(\mu)$ is established in § 11 is closely allied to the Wiener–Hopf solution of Chapter 5, which expresses the solution of the homogeneous Milne equation in terms of H-functions. The method of § 11 holds for a

† Chandrasekhar [2, 5] and [1], chap. v. Equation (8·1) was first obtained by Ambartsumian [1]. Following him, Russian writers use $\phi(\mu)$ in place of $H(\mu)$. The treatment in this chapter is mainly that of Busbridge [5].

‡ See (12·3) for the definition of ψ_n $(n = 0, 1, 2, \ldots)$.

wide class of functions $\Psi(x)$, including all polynomials. For a more general treatment reference should be made to Crum [1].

9. The characteristic equation

We shall assume that:

$\Psi(z)$ is real and non-negative for $0 \leqslant z \leqslant 1$; that every point of the closed interval $(-1, 1)$ lies in a domain D of the z-plane in which $\Psi(z)$ is regular; and that $\Psi(z)$ satisfies the condition (8·2).

Since every point of the interval $(-1, 1)$ is an interior point of D, $\Psi(z)$ is continuous in the interval and bounded in the neighbourhood of any point of the interval.

The function $T(\mu)$, defined by (8·4), can also be written (by means of (8·2))

$$T(\mu) = 1 - 2\psi_0 - 2\int_0^1 \frac{x^2 \Psi(x)}{\mu^2 - x^2}\, dx. \tag{9·1}$$

It is a many-valued function and we shall therefore denote by $T(\mu)$ the branch which is real when μ is real and $\mu > 1$. Then $T(\mu)$ is an even function of μ which is regular in the μ-plane cut along $(-1, 1)$. The equation

$$T(\mu) = 0 \tag{9·2}$$

is called the *characteristic equation.*

From (9·1) it is seen that

$$T(\mu) = 1 - 2\psi_0 + O(\mu^{-2}) \quad \text{as} \quad \mu \to \infty. \tag{9·3}$$

Since $\Psi(z)$ can be written in the form

$$\Psi(z) = \Psi_1(z^2) + z\Psi_2(z^2), \tag{9·4}$$

where each $\Psi_r(z^2)$ is an even function of z which is regular in D, therefore, for μ in D but not on the cut along $(-1, 1)$, $T(\mu)$ can be written

$$T(\mu) = 1 - 2\mu^2 \int_0^1 \{\Psi_1(x^2) - \Psi_1(\mu^2) + x[\Psi_2(x^2) - \Psi_2(\mu^2)]\} \frac{dx}{\mu^2 - x^2}$$
$$- \mu\Psi_1(\mu^2) \log\frac{\mu+1}{\mu-1} - \mu^2 \Psi_2(\mu^2) \log\frac{\mu^2}{\mu^2 - 1}, \tag{9·5}$$

where the logarithms are the principal values. The integral on the right of (9·5) represents a function regular in D with a double zero at $\mu = 0$, and it is therefore seen that

$$T(\mu) = 1 + O(|\mu|) \quad \text{as} \quad |\mu| \to 0 \tag{9·6}$$

uniformly in $\arg\mu$ in the cut plane.

9·1. *The roots of the characteristic equation.* Let μ_0 be a root of (9·2). If $\mu_0^2 = \xi + i\eta$, then, from (9·1),

$$1 - 2\psi_0 - 2\int_0^1 \frac{x^2\Psi(x)(\xi - x^2 - i\eta)}{(\xi - x^2)^2 + \eta^2}\,dx = 0.$$

Since $\Psi(x)$ is non-negative, on equating imaginary parts we have $\eta = 0$. Thus $\mu_0^2 = \xi$, where

$$2\int_0^1 \frac{x^2\Psi(x)}{\xi - x^2}\,dx = 1 - 2\psi_0. \tag{9·7}$$

Since μ_0 lies in the plane cut along $(-1, 1)$, either $\xi > 1$ and $\mu_0 = \pm\sqrt{\xi}$, or $\xi < 0$ and $\mu_0 = \pm i\sqrt{(-\xi)}$. But if $\xi < 0$ the left-hand side of (9·7) is negative (unless $\xi = -\infty$) and the right-hand side of (9·7) is non-negative; hence either ξ is infinite or $1 < \xi < \infty$.

When $\xi > 1$, the left-hand side of (9·7) is a steadily decreasing function of ξ and it tends to zero as $\xi \to \infty$; the right-hand side is positive or zero. When $\psi_0 = \frac{1}{2}$ there is one root $\xi = \infty$ and $T(\mu)$ has a double zero at infinity. When $\psi_0 < \frac{1}{2}$, $T(\infty)$ is positive and there will be one root ξ in $(1, \infty)$ provided that

$$\lim_{\mu \to 1+0} T(\mu) < 0, \tag{9·8}$$

and otherwise no root in $(1, \infty)$.

Thus there are three cases to consider:

(i) The conservative case, when $T(\mu)$ has a double zero at infinity;

(ii) The non-conservative case in which $\lim_{\mu \to 1+0} T(\mu) < 0$, when $T(\mu)$ has simple zeros at $\mu = \pm k^{-1}$ $(0 < k < 1)$;

(iii) The non-conservative case in which $\lim_{\mu \to 1+0} T(\mu) \geqslant 0$, when $T(\mu)$ has no zeros in the cut plane.

We shall refer to these cases by numbers.

From (9·4) and (9·5) it is seen that case (iii) can only arise when $\Psi(1) = 0$, for otherwise $T(\mu) \to -\infty$ as $\mu \to 1 + 0$.

10. The functional relation

We shall prove

THEOREM 10·1. *If $H(\mu)$ is any solution of* (8·1) *which is continuous in the interval* $0 \leqslant \mu \leqslant 1$,† *then* $1/H(\mu)$ *is regular in the* μ*-plane cut along* $(-1, 0)$; *and in the plane cut along* $(-1, 1)$, $H(\mu)$ *satisfies the functional relation*

$$H(\mu) H(-\mu) = 1/T(\mu). \tag{10·1}$$

In the plane cut along $(-1, 0)$, $H(\mu)$ *is regular except for a simple pole at infinity in case* (i) *and a simple pole at either* $1/k$ *or* $-1/k$ *in case* (ii).

The analytic properties of $1/H(\mu)$ follow at once from (8·1). Moreover, if (10·1) is true, the only zeros of $1/H(\mu)$ in the plane cut along $(-1, 0)$ must either be simple ones at the zeros of $T(\mu)$, or zeros in the interval $(0, 1)$. The latter do not exist because of the continuity of $H(\mu)$ in this interval, and the analytic properties of $H(\mu)$ therefore follow.

Let μ lie in the plane cut along $(-1, 1)$. Then, from (8·1),

$$\left\{1 - \frac{1}{H(\mu)}\right\}\left\{1 - \frac{1}{H(-\mu)}\right\} = \mu^2 \int_0^1 \int_0^1 \frac{\Psi(x) H(x) \Psi(y) H(y)}{(\mu+x)(\mu-y)} \, dx \, dy.$$

By means of the identity

$$\frac{\mu}{(\mu+x)(\mu-y)} = \frac{1}{x+y}\left\{\frac{y}{\mu-y} + \frac{x}{\mu+x}\right\}, \tag{10·2}$$

this can be written

$$\begin{aligned}\left\{1 - \frac{1}{H(\mu)}\right\}\left\{1 - \frac{1}{H(-\mu)}\right\} &= \mu \int_0^1 \frac{y\Psi(y) H(y)}{\mu - y} \, dy \int_0^1 \frac{\Psi(x) H(x)}{x+y} \, dx \\ &\quad + \mu \int_0^1 \frac{x\Psi(x) H(x)}{\mu+x} \, dx \int_0^1 \frac{\Psi(y) H(y)}{x+y} \, dy \\ &= \mu \int_0^1 \frac{\Psi(y) H(y)}{\mu - y}\left\{1 - \frac{1}{H(y)}\right\} dy \\ &\quad + \mu \int_0^1 \frac{\Psi(x) H(x)}{\mu + x}\left\{1 - \frac{1}{H(x)}\right\} dx,\end{aligned}$$

† By 'continuous in the interval $a \leqslant x \leqslant b$', we shall always mean 'continuous for $a < x < b$ and tending to finite limits as $x \to a+0$ and as $x \to b-0$'.

by (8·1). Again applying (8·1), we have

$$\left\{1-\frac{1}{H(\mu)}\right\}\left\{1-\frac{1}{H(-\mu)}\right\} = 1-\frac{1}{H(-\mu)}+1-\frac{1}{H(\mu)}$$
$$-\mu\int_0^1 \Psi(x)\left\{\frac{1}{\mu-x}+\frac{1}{\mu+x}\right\}dx$$
$$= 1-\frac{1}{H(-\mu)}-\frac{1}{H(\mu)}+T(\mu),$$

whence (10·1) follows.

11. The solution of the H-equation

The existence of $H(\mu)$ is proved in

THEOREM 11·1. *The integral equation* (8·1) *has a solution* $H(\mu)$, *which is regular in the* μ-*plane cut along* $(-1, 0)$ *except for a simple pole at infinity in case* (i) *and a simple pole at* $-k^{-1}$ *in case* (ii). *For* $\operatorname{re}\mu > 0$, $H(\mu)$ *is given by*†

$$H(\mu) = \exp\left\{\frac{\mu}{2\pi i}\int_{-i\infty}^{i\infty}\frac{\log T(z)}{z^2-\mu^2}dz\right\}. \tag{11·1}$$

In cases (i) *and* (iii) *this is the only solution of* (8·1) *which is continuous in the interval* $0 \leqslant \mu \leqslant 1$. *In case* (ii) *the function*

$$H_1(\mu) = \frac{1+k\mu}{1-k\mu}H(\mu) \tag{11·2}$$

is also such a solution, and there are no other solutions.

The integral (11·1) is known as Chandrasekhar's integral.

11·1. By Theorem 10·1, if $H(\mu)$ exists then it satisfies (10·1). Hence it must be possible to express $T(\mu)$ as the quotient of two functions, one of which, $1/H(-\mu)$, is regular outside the interval $(0, 1)$ and the other, $H(\mu)$, is regular outside the interval $(-1, 0)$ except, possibly, for a pole at $-k^{-1}$ (case (ii)) or at infinity (case (i)). The problem of such a representation is solved in a Wiener–Hopf solution of an integral equation and that method forms the basis of the following proof.

Let

$$\Phi(s) = \frac{s^2-1}{s^2-k^2}T(s^{-1}), \tag{11·3}$$

† $H(\mu)$ will always be used to denote this solution. Accurate tables of values for the case $\Psi(x) = \frac{1}{2}\omega_0$ $(0 < \omega_0 \leqslant 1)$ are given by Stibbs and Weir in [1].

where $k = 0$ in case (i), $\pm k^{-1}$ are the zeros of $T(\mu)$ in case (ii), and $k = 1$ in case (iii). By § 9·1, $\Phi(s)$ is regular and not zero in the s-plane cut along $(-\infty, -1)$ and $(1, \infty)$. In this cut plane, by (9·6),

$$\Phi(s) = 1+O(|s|^{-1}) \quad \text{as} \quad |s| \to \infty \tag{11·4}$$

uniformly in $\arg s$. From (11·3), (9·1) and (8·2), it is easily seen that $\Phi(it) > 0$ for $-\infty < t < \infty$ so that there is no variation in $\arg \Phi(s)$ along the imaginary axis. Hence $\log \Phi(s)$ can be defined to be that branch which is such that

$$\log \Phi(s) = O(|s|^{-1}) \quad \text{as} \quad |s| \to \infty \tag{11·5}$$

uniformly in $\arg s$. This branch is regular in the cut plane.

Let $0 < b < 1$ and let $|\operatorname{re} s| < b$. Let the z-plane be cut along $(-\infty, -1)$ and $(1, \infty)$, and let C be the rectangle with sides $\operatorname{re} z = \pm b$, $\operatorname{im} z = R$, $\operatorname{im} z = -R'$, R and R' being so large that $z = s$ lies within C. Then by Cauchy's integral theorem

$$\log \Phi(s) = \frac{1}{2\pi i}\int_C \frac{\log \Phi(z)}{z-s}\, dz.$$

By (11·5) the integrals along the sides $\operatorname{im} z = R$, $\operatorname{im} z = -R'$ tend to zero as $R \to \infty$, $R' \to \infty$, and therefore

$$\log \Phi(s) = \log \Phi^-(s) - \log \Phi^+(s), \tag{11·6}$$

where

$$\log \Phi^-(s) = \frac{1}{2\pi i}\int_{b-i\infty}^{b+i\infty} \frac{\log \Phi(z)}{z-s}\, dz, \tag{11·7}$$

$$\log \Phi^+(s) = \frac{1}{2\pi i}\int_{-b-i\infty}^{-b+i\infty} \frac{\log \Phi(z)}{z-s}\, dz, \tag{11·8}$$

each integral converging in virtue of (11·5). From (11·6),

$$\Phi(s) = \Phi^-(s)/\Phi^+(s), \tag{11·9}$$

and this will be shown to give the desired decomposition of $T(s^{-1})$.

Since, by (11·5), the integrand in (11·8) is $O(|z|^{-2})$ as $|z| \to \infty$, uniformly in $\arg z$, the line of integration can be deformed into a loop starting from infinity below the cut along $(-\infty, -1)$, passing round the branch point at -1 and returning to infinity above the cut. With the usual notation

$$\log \Phi^+(s) = \frac{1}{2\pi i}\int_{-\infty}^{(-1+)} \frac{\log \Phi(z)}{z-s}\, dz. \tag{11·10}$$

In particular, the contour can consist of the two sides of the cut and a small circle about $z = -1$. If s lies in any closed region of the s-plane cut along $(-\infty, -1)$, the integral (11·10) will converge uniformly, and hence $\log \Phi^+(s)$ is regular in this cut plane. It follows that $\Phi^+(s)$ is regular and not zero in the plane cut along $(-\infty, -1)$. Similarly, $\Phi^-(s)$ is regular and not zero in the plane cut along $(1, \infty)$.

Let $\operatorname{re} s < b$, and replace s by $-s$, z by $-z$ in (11·8). Then, since $\Phi(s)$ is an even function and by (11·7),

$$\log \Phi^+(-s) = -\log \Phi^-(s).$$

Hence
$$\Phi^+(-s)\,\Phi^-(s) = 1, \tag{11·11}$$

and this holds, by analytic continuation, in the plane cut along $(1, \infty)$.

Consider $\Phi^+(s)$ for large $|s|$. Let $s = \rho \exp i\phi$, where $|\phi| \leqslant \frac{1}{2}\pi$, and take the contour in (11·10) to be the sides of the cut and a small circle of radius η about $z = -1$. On the circle (by (9·5))

$$T(s^{-1}) = O(\ln \eta) \quad \text{and} \quad \log \Phi(s) = O(\ln|\ln \eta|).\dagger$$

Hence the integral round the circle is $O(\eta \ln|\ln \eta|) = o(1)$ as $\eta \to 0$. On the sides of the cut, $z = -x$ and

$$|z-s| \geqslant \rho, \quad |z-s| \geqslant x.$$

Hence, on choosing η sufficiently small and then using (11·5), we have

$$\begin{aligned} |\log \Phi^+(s)| &\leqslant O\left(\frac{1}{\rho}\int_{1+\eta}^{\rho} \frac{dx}{x}\right) + O\left(\int_{\rho}^{\infty} \frac{dx}{x^2}\right) + o(1) \\ &= O(\rho^{-1}\ln \rho) + O(\rho^{-1}) + o(1) \\ &= o(1) \quad \text{as} \quad \rho \to \infty. \end{aligned}$$

Thus
$$\Phi^+(s) \to 1 \quad \text{as} \quad |s| \to \infty \quad (|\arg s| \leqslant \tfrac{1}{2}\pi),$$

the convergence being uniform in $\arg s$. From (11·5), (11·9) and (11·11) it now follows that $\Phi^+(-s) \to 1$ as $|s| \to \infty$ uniformly for $|\arg s| \leqslant \frac{1}{2}\pi$, the plane being cut along $(1, \infty)$, and hence that

$$\Phi^+(s) \to 1 \quad \text{as} \quad |s| \to \infty \tag{11·12}$$

† ln will be used throughout for the natural logarithm.

uniformly in $\arg s$ in the plane cut along $(-\infty, -1)$. Also, by (11·11),

$$\Phi^-(s) \to 1 \quad \text{as} \quad |s| \to \infty \tag{11·13}$$

uniformly in $\arg s$ in the plane cut along $(1, \infty)$.

11·2. With the same convention about k, define

$$H(\mu) = \frac{1+\mu}{1+k\mu} \Phi^+(\mu^{-1}). \tag{11·14}$$

Then $H(\mu)$ is regular and not zero in the μ-plane cut along $(-1, 0)$ except for a simple pole at infinity in case (i) ($k = 0$), and a simple pole at $-k^{-1}$ in case (ii) ($0 < k < 1$). Also $1/H(\mu)$ is regular everywhere in the cut plane. By (11·12),

$$H(\mu) \to 1 \quad \text{as} \quad |\mu| \to 0, \tag{11·15}$$

the convergence being uniform in $\arg \mu$. From (11·14), (11·11), (11·9) and (11·3), it follows that $H(\mu)$ satisfies the functional relation (10·1).

If $\operatorname{re} s > 0$, we can take $b = 0$ in (11·8). Since $\Phi(z)$ is an even function of z,

$$\begin{aligned} \log \Phi^+(s) &= \frac{1}{2\pi i} \int_0^{i\infty} \log \Phi(z) \left\{ \frac{1}{z-s} - \frac{1}{z+s} \right\} dz \\ &= \frac{s}{\pi i} \int_0^{i\infty} \frac{\log \Phi(z)}{z^2 - s^2} dz. \end{aligned}$$

Put $s = \mu^{-1}$, $z = \zeta^{-1}$; then

$$\begin{aligned} \log \Phi^+(\mu^{-1}) &= \frac{\mu}{\pi i} \int_{-i\infty}^{0} \frac{\log \Phi(\zeta^{-1})}{\zeta^2 - \mu^2} d\zeta \\ &= \frac{\mu}{2\pi i} \int_{-i\infty}^{i\infty} \frac{\log \Phi(\zeta^{-1})}{\zeta^2 - \mu^2} d\zeta. \end{aligned} \tag{11·16}$$

In case (iii), $k = 1$, $\Phi(\zeta^{-1}) = T(\zeta)$, $\Phi^+(\mu^{-1}) = H(\mu)$, and (11·16) is (11·1). In the other two cases it can be reduced to (11·1) as follows:

Let $a > 0$, $\operatorname{re} \mu > 0$; then by integrating

$$[\mu \log(1 + a\zeta)]/(\zeta^2 - \mu^2)$$

round a semicircle on the right of the imaginary axis, we get

$$\frac{\mu}{2\pi i} \int_{-i\infty}^{i\infty} \frac{\log(1 + a\zeta)}{\zeta^2 - \mu^2} d\zeta = -\tfrac{1}{2} \log(1 + a\mu). \tag{11·17}$$

Similarly, by integrating

$$[\mu \log (1 - a\zeta)]/(\zeta^2 - \mu^2)$$

round a semicircle on the left, we get

$$\frac{\mu}{2\pi i}\int_{-i\infty}^{i\infty} \frac{\log (1 - a\zeta)}{\zeta^2 - \mu^2}\, d\zeta = -\tfrac{1}{2}\log (1 + a\mu). \tag{11·18}$$

Hence, by addition,

$$\frac{\mu}{2\pi i}\int_{-i\infty}^{i\infty} \frac{\log (1 - a^2\zeta^2)}{\zeta^2 - \mu^2}\, d\zeta = -\log (1 + a\mu) \quad (a > 0). \tag{11·19}$$

If we take $a = 1$ and subtract (11·19) from (11·16) (with $k = 0$), we get (11·1) in case (i). In case (ii) we have to add to (11·16), (11·19) with $a = k$, and then subtract (11·19) with $a = 1$.

11·3. We have next to prove that $H(\mu)$ satisfies (8·1). Let μ be any point of the μ-plane cut along $(-1, 0)$. Then $-\mu$ is a point of the z-plane cut along $(0, 1)$. From (11·14) and the fact that $\Phi^+(0)$ is finite and not zero, it follows that

$$1/H(z) \to 0 \quad \text{as} \quad z \to \infty \text{ (case (i))},$$

$$1/H(z) \to \text{a finite limit} \quad \text{as} \quad z \to \infty \text{ (cases (ii) and (iii))}.$$

Hence
$$\int_{|z|=R} \frac{dz}{z(z+\mu)\,H(-z)} \to 0 \quad \text{as} \quad R \to \infty. \tag{11·20}$$

Let Γ be a closed contour in the z-plane, cut along $(0, 1)$, which surrounds the cut but does not contain the point $-\mu$. On using (11·20), we get

$$\frac{1}{2\pi i}\int_{\Gamma} \frac{dz}{z(z+\mu)\,H(-z)} = \frac{1}{\mu H(\mu)}. \tag{11·21}$$

Let Γ consist of (i) the circle γ_0: $z = \epsilon \exp i\phi$, where ϕ runs from 0 to 2π, (ii) the lower side λ of the cut from $x = \epsilon$ to $x = 1 - \epsilon$, (iii) the circle γ_1: $z = 1 + \epsilon \exp i\phi$, where ϕ runs from $-\pi$ to π, (iv) the upper side λ' of the cut from $x = 1 - \epsilon$ to $x = \epsilon$. For sufficiently small ϵ, Γ is in D and $-\mu$ is outside Γ. By (10·1), (11·21) can be written

$$\frac{1}{H(\mu)} = \frac{\mu}{2\pi i}\int_{\gamma_0} \frac{dz}{z(z+\mu)\,H(-z)} + \frac{\mu}{2\pi i}\int_{\lambda \cup \gamma_1 \cup \lambda'} \frac{H(z)\,T(z)}{z(z+\mu)}\, dz. \tag{11·22}$$

By (8·4)

$$T(z) = 1 - 2z^2 \int_0^1 \frac{\Psi(y) - \Psi(z)}{z^2 - y^2} dy - z\Psi(z) \log \frac{z+1}{z-1}, \qquad (11·23)$$

where the logarithm is real when $z > 1$. Let

$$T_c(x) = 1 - 2x^2 \int_0^1 \frac{\Psi(y)}{x^2 - y^2} dy \quad (0 < x < 1), \qquad (11·24)$$

where the integral is evaluated as a Cauchy principal value. Then

$$T_c(x) = 1 - 2x^2 \int_0^1 \frac{\Psi(y) - \Psi(x)}{x^2 - y^2} dy - x\Psi(x) \ln \frac{1+x}{1-x}. \qquad (11·25)$$

From (11·23) and (11·25) it is seen that

$$\left.\begin{aligned} T(z) &= T_c(x) - i\pi x\Psi(x) \quad \text{on} \quad \lambda, \\ T(z) &= T_c(x) + i\pi x\Psi(x) \quad \text{on} \quad \lambda'. \end{aligned}\right\} \qquad (11·26)$$

By (11·15), the contribution of γ_0 to (11·22) is $1 + o(1)$ as $\epsilon \to 0$; and by (9·5) that of γ_1 is $O(\epsilon \ln \epsilon) = o(1)$. Hence, on letting $\epsilon \to 0$, (11·22) becomes

$$\frac{1}{H(\mu)} = 1 + \frac{\mu}{2\pi i} \int_0^1 \frac{H(x)}{x(x+\mu)} \{T_c(x) - i\pi x\Psi(x)$$

$$- T_c(x) - i\pi x\Psi(x)\} \, dx,$$

i.e.

$$\frac{1}{H(\mu)} = 1 - \mu \int_0^1 \frac{H(x)\Psi(x)}{x+\mu} dx,$$

and this is (8·1).

11·4. We come finally to the consideration of the uniqueness of the solution. Let $H_1(\mu)$ be any solution of (8·1) which is continuous in the interval $0 \leqslant \mu \leqslant 1$. By Theorem 10·1, $1/H_1(\mu)$ is regular in the μ-plane cut along $(-1, 0)$; it may have a simple zero at either $\mu = k^{-1}$ or at $\mu = -k^{-1}$ (case (ii)) and it satisfies the equation (10·1). Hence

$$H_1(\mu)\, H_1(-\mu) = 1/T(\mu) = H(\mu)\, H(-\mu), \qquad (11·27)$$

where $H(\mu)$ is the solution defined above.

Let

$$\phi(\mu) = H(\mu)/H_1(\mu). \qquad (11·28)$$

Then $\phi(\mu)$ is regular in the plane cut along $(-1, 0)$ except, possibly, for a simple pole at infinity (case (i)) or a simple pole at $\mu = -k^{-1}$ in case (ii). Since $H(\mu)$ is non-zero, the only possible zero of $\phi(\mu)$ is at $\mu = k^{-1}$ (case (ii)). From (11·27) we have

$$\phi(\mu)\,\phi(-\mu) = 1, \tag{11·29}$$

and it follows from this that $\phi(\mu)$ is regular in the whole μ-plane except, possibly, for a simple pole at infinity (case (i)) or at $-k^{-1}$ (case (ii)) and an isolated singularity at $\mu = 0$. However, by letting $\mu \to \infty$ in (11·29), it is seen that

$$\phi(\mu) \to \pm 1 \quad \text{as} \quad \mu \to \infty, \tag{11·30}$$

and therefore $\phi(\mu)$ cannot have a pole at infinity.

Consider $H_1(\mu)$ as $|\mu| \to 0$. If $\operatorname{re}\mu \geqslant 0$ and $0 \leqslant x \leqslant 1$, then $\mu + x| \geqslant |\mu|$ and $|\mu + x| \geqslant x$. It follows from (8·1) that, if $0 < \eta < 1$,

$$\left|\frac{1}{H_1(\mu)} - 1\right| \leqslant \int_0^\eta |H_1(x)|\,\Psi(x)\,dx + |\mu| \int_\eta^1 |H_1(x)|\,\Psi(x)\,\frac{dx}{x}$$

$$= o(1) \quad \text{as} \quad |\mu| \to 0$$

by choosing η (small) first and then μ. Hence

$$H_1(\mu) \to 1 \quad \text{as} \quad |\mu| \to 0 \quad (\operatorname{re}\mu \geqslant 0).$$

By (11·15), the same is true of $H(\mu)$ and therefore of $\phi(\mu)$. From (11·29) it now follows that

$$\phi(\mu) \to 1 \quad \text{as} \quad \mu \to 0 \tag{11·31}$$

in any manner. Hence $\phi(\mu)$ is regular at $\mu = 0$.

In cases (i) and (iii), $\phi(\mu)$ is a bounded integral function and it is therefore a constant, which must be unity. Thus the solution $H(\mu)$ is unique.

In case (ii), if $\phi(\mu)$ has a simple pole at $\mu = -k^{-1}$, it must also have a simple zero at $\mu = k^{-1}$ (by (11·29)). The only function with this pole and zero, regular elsewhere, and satisfying (11·30) and (11·31) is

$$\phi(\mu) = (1 - k\mu)/(1 + k\mu).$$

Hence the only other solution is

$$H_1(\mu) = H(\mu)\,(1 + k\mu)/(1 - k\mu). \tag{11·32}$$

It can easily be verified that this is a solution of (8·1) in case (ii). Since $\mu = -k^{-1}$ is a pole of $H(\mu)$, (8·1) gives

$$0 = 1 - \int_0^1 \frac{\Psi(x)\,H(x)}{1-kx}\,dx. \tag{11·33}$$

Hence

$$1-\mu\int_0^1 \frac{H_1(x)\,\Psi(x)}{\mu+x}\,dx = \int_0^1 \frac{\Psi(x)\,H(x)}{1-kx}\,dx$$

$$-\int_0^1 \frac{\mu(1+kx)}{(\mu+x)\,(1-kx)}\,H(x)\,\Psi(x)\,dx$$

$$= (1-k\mu)\int_0^1 \frac{xH(x)\,\Psi(x)}{(\mu+x)\,(1-kx)}\,dx. \tag{11·34}$$

But from (8·1) and (11·33)

$$\frac{1}{H(\mu)} = \int_0^1 \frac{\Psi(x)\,H(x)}{1-kx}\,dx - \mu\int_0^1 \frac{\Psi(x)\,H(x)}{\mu+x}\,dx$$

$$= (1+k\mu)\int_0^1 \frac{xH(x)\,\Psi(x)}{(\mu+x)\,(1-kx)}\,dx, \tag{11·35}$$

and hence

$$1-\mu\int_0^1 \frac{H_1(x)\,\Psi(x)}{\mu+x}\,dx = \frac{1-k\mu}{1+k\mu}\,\frac{1}{H(\mu)} = \frac{1}{H_1(\mu)}. \tag{11·36}$$

This completes the proof.

Equations (11·33) and (11·35) hold also when $k = 0$ (case (i)). Then (11·35) becomes

$$\frac{1}{H(\mu)} = \int_0^1 \frac{xH(x)\,\Psi(x)}{\mu+x}\,dx, \tag{11·37}$$

an important form of the H-equation in the conservative case.

12. The moments of $H(\mu)$

We shall define moments of two kinds:

$$\alpha_n = \int_0^1 H(\mu)\,\mu^n\,d\mu \quad (n = 0, 1, 2, \ldots), \tag{12·1}$$

$$h_n = \int_0^1 \Psi(\mu)\,H(\mu)\,\mu^n\,d\mu \quad (n = 0, 1, 2, \ldots). \tag{12·2}$$

We shall also write

$$\psi_n = \int_0^1 \Psi(\mu)\,\mu^n\,d\mu \quad (n = 0, 1, 2, \ldots). \tag{12·3}$$

THEOREM 12·1. *The moments of $H(\mu)$ satisfy the equations*†

$$h_0 = 1 - (1-2\psi_0)^{\frac{1}{2}}, \tag{12·4}$$

$$h_{2n}(1-2\psi_0)^{\frac{1}{2}} = \psi_{2n} - \tfrac{1}{2}(h_1 h_{2n-1} - h_2 h_{2n-2} + \ldots + h_{2n-1} h_1) \quad (n \geqslant 1). \tag{12·5}$$

To prove these, write (8·1) in the form

$$H(\mu) = 1 + \mu H(\mu) \int_0^1 \frac{\Psi(x)\,H(x)}{\mu + x}\,dx, \tag{12·6}$$

multiply by $\mu^{2n}\Psi(\mu)$ $(n \geqslant 0)$ and integrate with respect to μ over $(0, 1)$. Then

$$h_{2n} = \psi_{2n} + \int_0^1\int_0^1 \Psi(\mu)\,H(\mu)\,\Psi(x)\,H(x)\,\frac{\mu^{2n+1}}{\mu+x}\,d\mu\,dx.$$

On interchanging μ and x and adding the equations, we get

$$\begin{aligned} 2h_{2n} &= 2\psi_{2n} + \int_0^1\int_0^1 \Psi(\mu)\,H(\mu)\,\Psi(x)\,H(x) \\ &\qquad \times \{\mu^{2n} - \mu^{2n-1}x + \ldots + (-1)^r \mu^{2n-r}x^r + \ldots + x^{2n}\}\,d\mu\,dx \\ &= 2\psi_{2n} + h_{2n}h_0 - \ldots + (-1)^r h_{2n-r}h_r + \ldots + h_0 h_{2n}. \end{aligned} \tag{12·7}$$

If $n = 0$ this is $\qquad 2h_0 = 2\psi_0 + h_0^2,$

giving $\qquad h_0 = 1 \pm (1-2\psi_0)^{\frac{1}{2}}. \qquad (12·8)$

The sign is decided by considering the limit of $H(\mu)$ as $\mu \to \infty$. From (8·1),

$$1/H(\mu) \to 1 - h_0 \quad \text{as} \quad \mu \to \infty. \tag{12·9}$$

From (11·1), since $T(iy) > 0$,

$$H(\mu) = \exp\left\{-\frac{\mu}{2\pi}\int_{-\infty}^{\infty} \frac{\ln T(iy)}{\mu^2 + y^2}\,dy\right\} \quad (\mu > 0), \tag{12·10}$$

and therefore $H(\mu) \geqslant 0$ for $\mu > 0$. Hence $h_0 \leqslant 1$ and we must take the negative sign in (12·8). This proves (12·4), and (12·5) follows from (12·4) and (12·7).

† See Chandrasekhar [1], chap. v and Huang [1].

An immediate deduction from (12·4) is

COROLLARY 1. *The condition* (8·2) *is a necessary condition for* $H(\mu)$ *to be real for real values of* μ.

By writing (8·1) in the form

$$\frac{1}{H(\mu)} = 1 - \int_0^1 \Psi(x)\,H(x)\,dx + \int_0^1 \frac{x\Psi(x)\,H(x)}{\mu+x}\,dx$$

and using (12·4), we get

COROLLARY 2. *An alternative form for the H-equation is*

$$\frac{1}{H(\mu)} = (1-2\psi_0)^{\frac{1}{2}} + \int_0^1 \frac{x\Psi(x)\,H(x)}{\mu+x}\,dx. \tag{12·11}$$

When $\Psi(x) = \frac{1}{2}\omega_0$, then

$$h_n = \tfrac{1}{2}\omega_0\,\alpha_n \tag{12·12}$$

and Theorem 12·1 takes the important form

THEOREM 12·2. *When* $\Psi(x) = \frac{1}{2}\omega_0$, where $0<\omega_0\leqslant 1$, *the moments* α_n *satisfy the relations*

$$\tfrac{1}{2}\omega_0\,\alpha_0 = 1-(1-\omega_0)^{\frac{1}{2}}, \tag{12·13}$$

$$\alpha_{2n}(1-\omega_0)^{\frac{1}{2}} = (2n+1)^{-1} - \tfrac{1}{4}\omega_0(\alpha_1\,\alpha_{2n-1}-\alpha_2\,\alpha_{2n-2}+\ldots+\alpha_{2n-1}\,\alpha_1). \tag{12·14}$$

From this we have†

COROLLARY. *In the conservative case* ($\omega_0 = 1$)

$$\alpha_0 = 2 \quad \textit{and} \quad \alpha_1 = 2/\sqrt{3}. \tag{12·15}$$

When $\omega_0 \neq 1$, the odd moments have to be calculated numerically and then the even moments can be found from (12·14), but when $\omega_0 = 1$, the even moments have to be calculated and the odd moments deduced.

† When $n=1$, there is only one term in the series on the right-hand side of (12·14) and of (12·5).

Chapter 3

INTEGRAL OPERATORS

13. Introduction

Two integral operators, $\bar{\Lambda}$ and Λ, have already been defined (equations (4·9) and (5·14)). The solution of the Milne equation (5·13) involving Λ will later be shown to depend on the H-function with $\Psi(x) = \frac{1}{2}\omega_0$. The H-function with the general $\Psi(x)$ is obtained if $\frac{1}{2}\omega_0 E_1(t)$ is replaced by

$$K_1(t) = \int_1^\infty \Psi(x^{-1})\, x^{-1} \exp(-xt)\, dx. \tag{13·1}$$

Thus we define two new operators, L and $\bar{L}$, by the equations

$$L_\tau\{\phi(t)\} = \int_0^\infty \phi(t)\, K_1(|t-\tau|)\, dt, \tag{13·2}$$

$$\bar{L}_\tau\{\phi(t)\} = \int_0^{\tau_1} \phi(t)\, K_1(|t-\tau|)\, dt. \tag{13·3}$$

By considering, as we do in Part II, the integral equations

$$(1-L)_\tau\{\mathfrak{J}(t)\} = B(\tau), \tag{13·4}$$

$$(1-\bar{L})_\tau\{\mathfrak{J}(t)\} = B(\tau), \tag{13·5}$$

instead of (5·13) and (4·10), a wider range of problems is covered without any increase in the complexity of the mathematics. We shall call (13·4) and (13·5) the *generalized Milne equations.*

14. The functions $E_n(t)$ and $K_n(t)$

We shall always assume that $\Psi(x)$ has the properties italicized in § 9 and that

$$\Psi(0) > 0, \quad \Psi(1) > 0. \tag{14·1}$$

Generalizing (13·1), we define, for $n = 0, 1, 2, \ldots,$

$$K_n(t) = \int_1^\infty \Psi(x^{-1})\, x^{-n} \exp(-xt)\, dx. \tag{14·2}$$

The following properties of $E_n(\tau)$ are well known (see the appendix to Kourganoff [1]):

$$\left.\begin{aligned} E_n'(t) &= -E_{n-1}(t) \quad (n \geqslant 2, t > 0), \\ E_1'(t) &= -t^{-1}\exp(-t) \quad (t > 0), \end{aligned}\right\} \tag{14·3}$$

$$E_n(0) = 1/(n-1) \quad (n \geqslant 2), \tag{14·4}$$

$$E_1(t) = \ln t^{-1} + O(1) \quad (t \to +0), \tag{14·5}$$

$$E_n(t) \sim t^{-1}\exp(-t) \quad (t \to \infty). \tag{14·6}$$

The corresponding results for $K_n(t)$ are

$$K_n'(t) = -K_{n-1}(t) \quad (n \geqslant 1, t > 0), \tag{14·7}$$

$$K_n(0) = \psi_{n-2} \quad (n \geqslant 2), \tag{14·8}$$

$$K_1(t) = \Psi(0)\ln t^{-1} + O(1) \quad (t \to +0), \tag{14·9}$$

$$K_n(t) \sim \Psi(1)\, t^{-1}\exp(-t) \quad (t \to \infty). \tag{14·10}$$

In (14·8), ψ_{n-2} is the moment defined by (12·3). Since $\Psi(0) > 0$ and $\Psi(1) > 0$, the behaviour of $K_n(t)$ for small and large t is similar to that of $E_n(t)$.

The proofs of (14·7) and (14·8) are trivial. To prove (14·9), put $xt = u$ in (13·1); then

$$\begin{aligned} K_1(t) - \Psi(0)\ln t^{-1} &= \int_1^\infty \Psi(tu^{-1})\exp(-u)\frac{du}{u} \\ &\quad + \int_t^1 \{\Psi(tu^{-1}) - \Psi(0)\}\exp(-u)\frac{du}{u} \\ &\quad - \Psi(0)\int_t^1 \{1 - \exp(-u)\}\frac{du}{u}. \end{aligned}$$

The first and third integrals are clearly bounded as $t \to +0$. Since $\Psi(x)$ has a continuous derivative for $0 \leqslant x \leqslant 1$, the second integral is

$$O\left(t\int_t^1 u^{-2}\,du\right) = O(1) \quad \text{as} \quad t \to +0.$$

Hence we have (14·9).

On integrating (14·2) by parts, we get

$$|t\{\exp t K_n(t) - \Psi(1)\, t^{-1}\}|$$

$$= \left|\exp t \int_1^\infty \exp(-xt)\,\{n\Psi(x^{-1})\,x^{-n-1} + \Psi'(x^{-1})\,x^{-n-2}\}\,dx\right|$$

$$= O\left(\exp t \int_1^\infty \exp(-xt)\,dx\right)$$

$$= O(t^{-1})$$

$$\to 0 \quad \text{as} \quad t \to \infty.$$

Hence (14·10) follows.

15. Properties of the operators L and $\bar{L}$

The general properties of L and $\bar{L}$ are similar. We shall work with L, stating the corresponding results for $\bar{L}$ only where these differ.

15·1. *Positivity.* Since the kernel of L is positive, if $\phi(\tau) \geqslant 0$, then $L_\tau\{\phi\} \geqslant 0$ for $\tau \geqslant 0$, and there cannot be equality for any τ unless $\phi = 0$ for almost all $\tau \geqslant 0$. In particular, *if $\phi \leqslant \chi$ and if $\phi \neq \chi$ in a set of positive measure, then*

$$L_\tau\{\phi\} < L_\tau\{\chi\}. \tag{15·1}$$

If $|\phi| \leqslant \chi$, then it is seen from (13·2) that

$$|L_\tau\{\phi\}| \leqslant L_\tau\{|\phi|\} \leqslant L_\tau\{\chi\}. \tag{15·2}$$

On using (14·7), (14·8) and (14·10), we get

$$L_\tau\{1\} = 2\psi_0 - K_2(\tau) \tag{15·3}$$

and therefore

$$L_\tau\{1\} \leqslant 2\psi_0. \tag{15·4}$$

Hence, *if $|\phi(\tau)| \leqslant M$ for $\tau \geqslant 0$, where M is a constant, then*

$$|L_\tau\{\phi\}| \leqslant 2M\psi_0. \tag{15·5}$$

Let L^n_τ denote the operator L repeated n times, L^0_τ being the identity operator. Then repeating the argument, we have

$$|L^n_\tau\{\phi\}| \leqslant M(2\psi_0)^n \quad (n = 0, 1, 2, \ldots). \tag{15·6}$$

The above inequalities all hold when L is replaced by $\bar{L}$. The proofs of (15·5) and (15·6) follow from the inequality

$$\bar{L}_\tau\{1\} \leqslant L_\tau\{1\} \leqslant 2\psi_0. \tag{15·7}$$

15·2. *Linearity.* L and $\mathbf{L}$ are linear in the sense that, if a and b are constants,

$$L_\tau\{a\phi_1(t)+b\phi_2(t)\} = aL_\tau\{\phi_1(t)\}+bL_\tau\{\phi_2(t)\}.$$

Hence if $\mathfrak{J}_1(\tau)$, $\mathfrak{J}_2(\tau)$ are solutions of

$$(1-L)_\tau\{\mathfrak{J}_r(t)\} = B_r(\tau) \quad (r = 1, 2), \tag{15·8}$$

where the functions $B_r(\tau)$ are known, then $\mathfrak{J}(\tau) = a\mathfrak{J}_1(\tau)+b\mathfrak{J}_2(\tau)$ is a solution of

$$(1-L)_\tau\{\mathfrak{J}(t)\} = aB_1(\tau)+bB_2(\tau). \tag{15·9}$$

Thus solutions of Milne equations can be combined linearly to give solutions of more complicated equations. Moreover, if corresponding lower-case letters denote Laplace transforms (a notation which we shall always use), then

$$\mathfrak{j}(s) = a\mathfrak{j}_1(s)+b\mathfrak{j}_2(s). \tag{15·10}$$

Thus the corresponding 'emergent intensities' (see (5·15)) are combined in the same way to give the new emergent intensity.

15·3. *Smoothing effect.*

THEOREM 15·1. *If $\phi(t)$ is measurable and bounded in every interval (α, β) such that $0<\alpha<\beta<\infty$, and if*

$$\left.\begin{aligned}\phi(t) &= O(\ln t^{-1}) \quad \text{as} \quad t\to+0,\\ \phi(t) &= O[\exp(at)] \quad \text{as} \quad t\to\infty,\end{aligned}\right\} \tag{15·11}$$

where $0\leqslant a<1$, then $L_\tau\{\phi(t)\}$ exists for $\tau\geqslant 0$, is continuous for $\tau>0$, and

$$\left.\begin{aligned}L_\tau\{\phi(t)\} &\to L_0\{\phi(t)\} \quad \text{as} \quad \tau\to+0,\\ L_\tau\{\phi(t)\} &= O[\exp(a\tau)] \quad \text{as} \quad \tau\to\infty.\end{aligned}\right\} \tag{15·12}$$

In (15·11) *and* (15·12), $\exp(a\tau)$ *may be replaced by* τ^α $(\alpha\geqslant 0)$.

The existence of $L_\tau\{\phi(t)\}$ for $\tau\geqslant 0$ follows at once from (14·9), (14·10) and the conditions of the theorem.

Since the smoothing effect of an integral operator is well known, a proof will only be given of (15·12). When $\exp(a\tau)$ is replaced by τ^α, the necessary changes in the proof are easily made.

Let $\epsilon > 0$. By (14·10) we can choose $\beta > 1$ such that, when $0 \leqslant \tau < \frac{1}{2}$,

$$\left| \int_\beta^\infty \phi(t) K_1(t-\tau)\, dt \right| = O\left(\int_\beta^\infty \exp\left[-(1-a)t \right] dt \right) < \tfrac{1}{6}\epsilon.$$

Hence, if $0 < \tau < \frac{1}{2}$,

$$| L_\tau\{\phi\} - L_0\{\phi\} | < \tfrac{1}{3}\epsilon + \int_0^\beta | \phi(t) | . | K_1(| t-\tau |) - K_1(t) | \, dt$$

$$< \tfrac{2}{3}\epsilon + \int_0^1 | \phi(t) | . | K_1(| t-\tau |) - K_1(t) | \, dt$$

if $\tau < \tau_1(\epsilon)$, by the uniform continuity of $K_1(t)$ for $\frac{1}{2} \leqslant t \leqslant \beta$. By (15·11) and (14·9)

$$| L_\tau\{\phi\} - L_0\{\phi\} | < \tfrac{2}{3}\epsilon + O\left(\int_0^\tau \ln t^{-1} \ln (\tau - t)^{-1} dt \right) + O\left(\int_0^\tau [\ln t^{-1}]^2 dt \right)$$

$$+ O\left(\int_\tau^1 \ln t^{-1} [K_1(t-\tau) - K_1(t)]\, dt \right). \qquad (15·13)$$

The first integral on the right is

$$2\int_0^{\frac{1}{2}\tau} \ln t^{-1} \ln (\tau - t)^{-1} dt \leqslant 2 \ln (\tfrac{1}{2}\tau)^{-1} \int_0^{\frac{1}{2}\tau} \ln t^{-1} dt = O[\tau (\ln \tau^{-1})^2].$$

The third integral is less than

$$\ln \tau^{-1} \int_\tau^1 [K_1(t-\tau) - K_1(t)]\, dt$$

$$= \ln \tau^{-1} [K_2(0) - K_2(\tau) + K_2(1) - K_2(1-\tau)]$$

by (14·7). But, by (14·7) and (14·9),

$$K_2(0) - K_2(\tau) = \Psi(0)\, \tau \ln \tau^{-1} + O(\tau),$$

$$K_2(1) - K_2(1-\tau) = O(\tau),$$

and therefore the third integral in (15·13) is $O[\tau(\ln \tau^{-1})^2]$. On evaluating the second integral, we get

$$| L_\tau\{\phi\} - L_0\{\phi\} | < \tfrac{2}{3}\epsilon + O[\tau(\ln \tau^{-1})^2] < \epsilon$$

if $\tau < \tau_0(\epsilon)$. This proves the first part of (15·12).

For large τ, by (15·11) and (14·10),

$$\left|\int_0^\tau \phi(t)\,K_1(\tau-t)\,dt\right| = \left|\int_0^\tau \phi(\tau-u)\,K_1(u)\,du\right|$$

$$= O\left[\exp(a\tau)\int_0^{\tau-1}\exp(-au)\,K_1(u)\,du\right]$$

$$+O\left[\exp(-\tau)\int_{\tau-1}^{\tau}|\phi(\tau-u)|\,du\right]$$

$$= O\left[\exp(a\tau)\int_0^{\infty}\exp(-au)\,K_1(u)\,du\right]$$

$$+O\left[\exp(-\tau)\int_0^1|\phi(t)|\,dt\right]$$

$$= O[\exp(a\tau)]$$

because $a \geqslant 0$. Also

$$\left|\int_\tau^\infty \phi(t)\,K_1(t-\tau)\,dt\right| = \left|\int_0^\infty \phi(\tau+u)\,K_1(u)\,du\right|$$

$$= O\left[\exp(a\tau)\int_0^{\infty}\exp(au)\,K_1(u)\,du\right]$$

$$= O[\exp(a\tau)]$$

since $a < 1$. Hence

$$L_\tau\{\phi\} = O[\exp(a\tau)] \quad \text{as} \quad \tau\to\infty.$$

For $\bar{L}$ we have, similarly,

THEOREM 15·2. *If $\phi(t)$ is measurable and bounded in every interval (α,β) such that $0<\alpha<\beta<\tau_1$ and if*

$$\left.\begin{aligned}\phi(t) &= O[\ln t^{-1}] \quad \textit{as} \quad t\to+0,\\ \phi(t) &= O[\ln(\tau_1-t)^{-1}] \quad \textit{as} \quad t\to\tau_1-0,\end{aligned}\right\} \tag{15·14}$$

then $\bar{L}_\tau\{\phi(t)\}$ exists for $0\leqslant\tau\leqslant\tau_1$, is continuous for $0<\tau<\tau_1$, and

$$\left.\begin{aligned}\bar{L}_\tau\{\phi(t)\}&\to\bar{L}_0\{\phi(t)\} \quad \textit{as} \quad \tau\to+0,\\ \bar{L}_\tau\{\phi(t)\}&\to\bar{L}_{\tau_1}\{\phi(t)\} \quad \textit{as} \quad \tau\to\tau_1-0.\end{aligned}\right\} \tag{15·15}$$

From this theorem it follows that $\bar{L}_\tau\{\phi(t)\}$ is continuous in the interval $0\leqslant\tau\leqslant\tau_1$ in the sense of the footnote on p. 16.

15·4. *The differentiation formula.*†

THEOREM 15·3. *If $\phi(t)$ is continuous for $t \geqslant 0$ and has a continuous derivative for $t > 0$ which is such that*

$$\left.\begin{aligned} \phi'(t) &= O(\ln t^{-1}) \quad \text{as} \quad t \to +0, \\ \phi'(t) &= O[\exp(at)] \quad \text{as} \quad t \to \infty, \end{aligned}\right\} \tag{15·16}$$

where $0 \leqslant a < 1$, then

$$\frac{d}{d\tau} L_\tau\{\phi(t)\} = L_\tau\{\phi'(t)\} + \phi(0)\, K_1(\tau). \tag{15·17}$$

Let

$$F(\tau) = L_\tau\{\phi'(t)\} + \phi(0)\, K_1(\tau). \tag{15·18}$$

By (15·16), Theorem 15·1, and the properties of $K_1(\tau)$, $F(\tau)$ is continuous for $\tau > 0$ and it is $O(\ln \tau^{-1})$ as $\tau \to +0$. Hence $\int_0^\tau F(t)\,dt$ exists. This is also true if $\phi'(t)$ is replaced by $|\phi'(t)|$ and the inversions of the orders of integration in the following analysis are therefore justified.

We can write $F(t)$ in the form

$$F(t) = \int_0^t \phi'(t-u)\, K_1(u)\, du + \int_0^\infty \phi'(t+u)\, K_1(u)\, du + \phi(0)\, K_1(t).$$

Hence, on integrating over $(0, \tau)$ and inverting the orders of integration, we have

$$\begin{aligned} \int_0^\tau F(t)\,dt &= \int_0^\tau K_1(u)\,du \int_u^\tau \phi'(t-u)\,dt \\ &\quad + \int_0^\infty K_1(u)\,du \int_0^\tau \phi'(t+u)\,dt + \phi(0) \int_0^\tau K_1(t)\,dt \\ &= \int_0^\tau K_1(u)\,[\phi(\tau-u) - \phi(0)]\,du \\ &\quad + \int_0^\infty K_1(u)\,[\phi(\tau+u) - \phi(u)]\,du + \phi(0) \int_0^\tau K_1(t)\,dt. \end{aligned} \tag{15·19}$$

By (15·16), $\phi(\tau)$ is $O[\exp(a\tau)]$ as $\tau \to \infty$ if $a > 0$, and $O(\tau)$ as $\tau \to \infty$ if $a = 0$. Hence (15·19) can be rewritten in the form

$$\int_0^\tau F(t)\,dt = L_\tau\{\phi(t)\} - \int_0^\infty K_1(u)\,\phi(u)\,du,$$

† Hopf [1], p. 34. The proof given here is simpler.

the integrals existing. Since $F(\tau)$ is continuous for $\tau > 0$,

$$F(\tau) = \frac{d}{d\tau} L_\tau\{\phi(t)\},$$

and this proves the theorem.

For $\bar{L}$ the corresponding theorem is

THEOREM 15·4. *If* $\phi(t)$ *is continuous for* $0 \leqslant t \leqslant \tau_1$ *and has a continuous derivative for* $0 < t < \tau_1$ *which is such that*

$$\left.\begin{aligned} \phi'(t) &= O(\ln t^{-1}) \quad as \quad t \to +0, \\ \phi'(t) &= O[\ln(\tau_1 - t)^{-1}] \quad as \quad t \to \tau_1 - 0, \end{aligned}\right\} \tag{15·20}$$

then

$$\frac{d}{d\tau} \bar{L}_\tau\{\phi(t)\} = \bar{L}_\tau\{\phi'(t)\} + \phi(0)\, K_1(\tau) - \phi(\tau_1)\, K_1(\tau_1 - \tau). \tag{15·21}$$

15·5. *Symmetry.* We shall write

$$(\phi, \chi) = \int_0^\infty \phi(t)\, \chi(t)\, dt. \tag{15·22}$$

Since the kernel of L is symmetrical in t and τ,

$$(\phi, L_\tau\{\chi\}) = (L_t\{\phi\}, \chi) = \int_0^\infty \int_0^\infty K_1(|t-\tau|)\, \phi(\tau)\, \chi(t)\, d\tau\, dt, \tag{15·23}$$

at least formally. The equation will be true whenever the integrals converge absolutely. It will also be true, if ϕ and χ are non-negative, in the sense that either all integrals are finite and equal or all are infinite. Equation (15·23) is known as the symmetrical property of L.

If the upper limits in (15·22) and (15·23) are replaced by τ_1, the symmetrical property is true also for $\bar{L}$.

PART II
MILNE EQUATIONS

CHAPTER 4

ITERATIVE SOLUTIONS

16. Introduction

Since $K_1(|t-\tau|)\to\infty$ as $t\to\tau$, the kernel of the operators L and $\bar{L}$ (see (13·2) and (13·3)) is singular. Nevertheless, solutions of the generalized Milne equations (13·4) and (13·5) can, in many cases, be found by iteration as for a bounded kernel.

Iterative solutions are of two kinds. When $\psi_0 = \frac{1}{2}$ (the conservative case) sequences of iterations can be found which converge to the solution of the homogeneous Milne equation (§ 17). This fact has proved of considerable practical importance. When $\psi_0 \leqslant \frac{1}{2}$, the non-homogeneous equation

$$(1-L)_\tau\{\mathfrak{J}(t)\} = B(\tau) \tag{16·1}$$

can be solved, at least formally, by iteration to give

$$\mathfrak{J}(\tau) = \sum_{\nu=0}^{\infty} L_\tau^\nu\{B(t)\}. \tag{16·2}$$

This is the Neumann series (N-series) of the equation. If it converges to a solution of (16·1), we shall call this the N-solution and denote it by $\mathfrak{J}_N(\tau)$. Most of this chapter, which is based on Hopf [1], Chapter II, is concerned with N-solutions.

17. The homogeneous Milne equation ($\psi_0 = \frac{1}{2}$)

The iterated sequences which converge to the solution are given in

THEOREM 17·1. *Let* $\psi_0 = \frac{1}{2}$ *and let*

$$f_0(\tau) = \tau + 2\psi_1, \quad g_0(\tau) = \tau + 1. \tag{17·1}$$

If $$f_0(\tau) \leqslant \phi_0(\tau) \leqslant g_0(\tau), \tag{17·2}$$
where $\phi_0(\tau)$ *is measurable over* $(0, \alpha)$ *for every finite* α, *then the sequences defined by*

$$f_n = L_\tau^n\{f_0\}, \quad \phi_n = L_\tau^n\{\phi_0\}, \quad g_n = L_\tau^n\{g_0\} \tag{17·3}$$

all converge, as $n \to \infty$, *to a function* $f(\tau)$, *which is a solution of*

$$(1-L)_\tau\{f(t)\} = 0. \tag{17·4}$$

The function $f(\tau)$ *is of the form*

$$f(\tau) = \tau + q(\tau), \quad 2\psi_1 < q(\tau) < 1. \tag{17·5}$$

By (14·7)–(14·10) and the definition of L, it is easily verified that, if c is a constant,

$$(1-L)_\tau\{t+c\} = cK_2(\tau) - K_3(\tau). \tag{17·6}$$

Now $$\frac{d}{d\tau}\frac{K_2(\tau)}{K_3(\tau)} = \frac{K_2^2(\tau) - K_1(\tau)K_3(\tau)}{K_3^2(\tau)},$$

and by Schwarz's inequality

$$K_2^2(\tau) = \left\{\int_1^\infty \Psi^{\frac{1}{2}}(x^{-1})\,x^{-\frac{1}{2}}\exp(-\tfrac{1}{2}x\tau)\,.\,\Psi^{\frac{1}{2}}(x^{-1})\,x^{-\frac{3}{2}}\exp(-\tfrac{1}{2}x\tau)\,dx\right\}^2$$
$$\leqslant \int_1^\infty \Psi(x^{-1})\,x^{-1}\exp(-x\tau)\,dx \int_1^\infty \Psi(x^{-1})\,x^{-3}\exp(-x\tau)\,dx$$
$$= K_1(\tau)\,K_3(\tau).$$

Hence $(d/d\tau)\{K_2(\tau)/K_3(\tau)\} \leqslant 0$. By considering the values of $K_2(\tau)/K_3(\tau)$ as $\tau \to \infty$ and as $\tau \to 0$, we have

$$1 \leqslant K_2(\tau)/K_3(\tau) \leqslant \psi_0/\psi_1 = 1/2\psi_1. \tag{17·7}$$

Hence, taking $c = 2\psi_1$ and $c = 1$ in (17·6),

$$(1-L)_\tau\{f_0\} \leqslant 0, \quad (1-L)_\tau\{g_0\} \geqslant 0,$$

i.e. by (17·3), $$f_0 \leqslant f_1, \quad g_0 \geqslant g_1, \tag{17·8}$$

and since $K_3(\tau)$ is not a constant multiple of $K_2(\tau)$, there is inequality in each case in a set of positive measure.† From (17·2) and (17·8), by repeated application of the operator L, it follows that

$$f_0 \leqslant f_1 < \ldots < f_n \leqslant \phi_n \leqslant g_n < \ldots < g_1 \leqslant g_0. \tag{17·9}$$

† Hence, on applying the operator L to (17·8), there is inequality for all τ (see § 15·1).

Hence, as $n \to \infty$,

$$f_n(\tau) \to f(\tau), \quad g_n(\tau) \to g(\tau)$$

such that

$$f_0 < f \leqslant g < g_0, \tag{17·10}$$

and therefore

$$0 \leqslant g - f < 1 - 2\psi_1. \tag{17·11}$$

Since the sequence (f_n) is monotonic,

$$\lim_{n\to\infty} f_{n+1} = \lim_{n\to\infty} L_\tau\{f_n\} = L_\tau\left\{\lim_{n\to\infty} f_n\right\},$$

and similarly for g_n. Hence

$$f = L_\tau\{f\}, \quad g = L_\tau\{g\}. \tag{17·12}$$

Thus $f(\tau)$ satisfies (17·4), and since it lies between $\tau + 2\psi_1$ and $\tau + 1$, it is of the form (17·5). By Theorem 15·1, $f(\tau)$ is continuous for $\tau \geqslant 0$.†

From (17·12) it follows that $g - f$ is also a solution of the homogeneous Milne equation. By (17·11) it is non-negative and bounded, and by Theorem 15·1 it is continuous for $\tau \geqslant 0$. In the first theorem of § 18 it will be shown that such a function is of the form $cf(\tau)$, where c is a constant, and since $g - f$ is bounded, $c = 0$. Assuming this, we have $f = g$, and on letting $n \to \infty$ in (17·9) it is seen that $\phi_n \to f$.

The function $q(\tau)$ is known as *Hopf's function*.

18. Uniqueness

The most precise uniqueness theorems for the homogeneous Milne equation will be obtained by the Wiener–Hopf method in the next chapter. The following theorem is a simplified version‡ of one given by Hopf in [1]:

THEOREM 18·1. *Let* $\psi_0 = \frac{1}{2}$; *if* $\phi(\tau)$ *is continuous and non-negative for* $\tau \geqslant 0$ *and if*

$$(1 - L)_\tau\{\phi(t)\} = 0, \tag{18·1}$$

then $\phi(\tau) = cf(\tau)$, *where* $f(\tau)$ *is the function of Theorem* 17·1 *and* c *is a constant.*

Since $f(\tau)$ is continuous and positive for $\tau \geqslant 0$, therefore $\phi(\tau)/f(\tau)$ is continuous and non-negative. Let c be its lower

† The continuity at $\tau = 0$ is from the right.
‡ This was suggested by E. C. Titchmarsh.

bound for $\tau \geqslant 0$. Then either (i) this is attained at a finite point τ', or (ii) $\phi(\tau)/f(\tau) \to c$ as $\tau \to \infty$ through some sequence of points. Let

$$\chi(\tau) = \phi(\tau) - cf(\tau). \tag{18·2}$$

Then $\chi(\tau) \geqslant 0$ and

$$(1-L)_\tau\{\chi(t)\} = 0, \tag{18·3}$$

and either (i) $\chi(\tau') = 0$, or (ii) $\chi(\tau)/\tau \to 0$ as $\tau \to \infty$ through some sequence (since $f(\tau) = \tau + O(1)$ for large τ).

In case (i), by (18·3) with $\tau = \tau'$,

$$L_{\tau'}\{\chi(t)\} = 0,$$

and hence (see § 15·1) $\chi(\tau) = 0$ since it is non-negative and continuous. Thus $\phi(\tau) = cf(\tau)$.

In case (ii), from (18·3),

$$\chi(\tau) = \int_0^\infty K_1(|t-\tau|)\,\chi(t)\,dt. \tag{18·4}$$

On integrating with respect to τ over $(0, x)$ and using the condition $\psi_0 = \frac{1}{2}$, we obtain [cf. (6·12) and (6·13)]

$$\int_0^\infty \chi(t)\,K_2(t)\,dt = \int_x^\infty \chi(t)\,K_2(t-x)\,dt - \int_0^x \chi(t)\,K_2(x-t)\,dt. \tag{18·5}$$

The existence of these integrals follows from the existence of that on the right of (18·4). The left-hand side of (18·5) is a constant C, say. On integrating with respect to x over $(0, \tau)$, we get [cf. (7·10)]

$$C\tau = \int_0^\infty \chi(t)\,K_3(|t-\tau|)\,dt - \int_0^\infty \chi(t)\,K_3(t)\,dt. \tag{18·6}$$

Now $K_3(u) \leqslant K_1(u)$ and hence, as $\tau \to \infty$,

$$C \leqslant \tau^{-1}\int_0^\infty \chi(t)\,K_1(|t-\tau|)\,dt + O(\tau^{-1}) = \tau^{-1}\chi(\tau) + O(\tau^{-1}).$$

Since this tends to zero as $\tau \to \infty$ through the sequence, it follows that $C = 0$, i.e.

$$\int_0^\infty \chi(t)\,K_2(t)\,dt = 0,$$

and hence $\chi(t) = 0$. Thus again $\phi(\tau) = cf(\tau)$.

COROLLARY. *When* $\psi_0 = \frac{1}{2}$, *every continuous solution of* (18·1) *with a finite lower bound is of the form* $cf(\tau)$.

If $\phi(\tau)$ is the solution, then $\phi(\tau)+af(\tau)$ is also a solution and the constant a can be chosen so that this is non-negative for $\tau \geqslant 0$. By the theorem, $\phi(\tau)+af(\tau)$ is a multiple of $f(\tau)$ and hence $\phi(\tau) = cf(\tau)$.

18·1. In the non-conservative case it is easy to prove

THEOREM 18·2. *When* $\psi_0 < \frac{1}{2}$, *there is no non-zero bounded solution of the homogeneous Milne equation.*

Let $\phi(\tau)$ be a solution such that $|\phi(\tau)| \leqslant M$. By repeated application of L, we have

$$\phi = L_\tau\{\phi\} = L_\tau^2\{\phi\} = \ldots = L_\tau^n\{\phi\},$$

and hence, by (15·6),

$$|\phi(\tau)| \leqslant M(2\psi_0)^n.$$

Since $(2\psi_0)^n \to 0$ as $n \to \infty$, $\phi(\tau) = 0$.

19. The non-homogeneous equation

We shall usually suppose that the function $B(\tau)$ in (16·1) is continuous for $\tau > 0$ (but not necessarily positive) and that

$$\left.\begin{aligned} B(\tau) &= O(\ln \tau^{-1}) \quad \text{as} \quad \tau \to +0, \\ B(\tau) &= O[\exp(a\tau)] \quad \text{as} \quad \tau \to \infty, \end{aligned}\right\} \tag{19·1}$$

where $0 \leqslant a < 1$. These conditions are sufficient to ensure the existence of $L_\tau^\nu\{B(t)\}$ for $\nu \geqslant 1$. By Theorem 15·1, each is continuous for $\tau > 0$, bounded as $\tau \to 0$ and $O[\exp(a\tau)]$ as $\tau \to \infty$.

It is convenient to introduce the notation

$$B(\tau) \in C(\ln \tau^{-1}, \exp a\tau),$$

the C denoting continuity for $\tau > 0$. The order condition at $\tau = 0$ will always be placed first and then that at infinity. With this notation $L_\tau^\nu\{B(t)\} \in C(1, \exp a\tau)$ for $\nu \geqslant 1$.

When a particular solution of (16·1) is known, the general solution is obtained by adding a multiple of $J(\tau)$, where $J(\tau)$ is the non-zero solution (if it exists in the class of functions considered) of the homogeneous equation. The particular solution will usually be the N-solution.

19·1. *The N-solution of* (16·1). Let

$$\mathfrak{J}_N^n(\tau) = \sum_{\nu=0}^{n} L_\tau^\nu\{B(t)\}. \tag{19·2}$$

When the N-series is convergent for $\tau > 0$, $\mathfrak{J}_N^n(\tau)$ tends, as $n \to \infty$, to a measurable function $\mathfrak{J}_N(\tau)$. From (19·2) we have

$$\mathfrak{J}_N^{n+1}(\tau) = L_\tau\{\mathfrak{J}_N^n(t)\} + B(\tau). \tag{19·3}$$

Hence $\mathfrak{J}_N(\tau)$ will be the N-solution provided that

$$\lim_{n\to\infty} L_\tau\{\mathfrak{J}_N^n(t)\} = L_\tau\left\{\lim_{n\to\infty} \mathfrak{J}_N^n(t)\right\}.$$

This is true if $B(\tau) \geqslant 0$, for then the sequence $\mathfrak{J}_N^n(\tau)$ $(n = 1, 2, \ldots)$ is monotonic for each fixed τ. Thus we have

THEOREM 19·1. *Let* $B(\tau) \geqslant 0$ *in* (16·1). *If the N-series* (16·2) *exists and is convergent for* $\tau > 0$, *its sum is the N-solution of the equation.*

Since
$$\mathfrak{J}_N(\tau) = L_\tau\{\mathfrak{J}_N(t)\} + B(\tau), \tag{19·4}$$

$\mathfrak{J}_N(\tau)$ will be continuous for $\tau > 0$ if $B(\tau)$ and $L_\tau\{\mathfrak{J}_N(t)\}$ are continuous. Sufficient conditions for this can be deduced from

THEOREM 19·2. *Let* $B(\tau)$ *be continuous for* $\tau > 0$ *and let* $\mathfrak{J}(\tau)$ *be any solution of* (16·1). *If* $\mathfrak{J}(\tau)$ *is bounded in every interval* $0 < \alpha < \beta < \infty$, *and if it is* $O(\ln \tau^{-1})$ *as* $\tau \to +0$ *and* $O[\exp(a\tau)]$ *as* $\tau \to \infty$ $(0 \leqslant a < 1)$, *then* $\mathfrak{J}(\tau) \in C(\ln \tau^{-1}, \exp a\tau)$. *In particular, if* $\mathfrak{J}(\tau)$ *is bounded for* $\tau \geqslant 0$, *then* $\mathfrak{J}(\tau) \in C(1, 1)$.

The existence of $L_\tau\{\mathfrak{J}(t)\}$ implies the measurability of $\mathfrak{J}(\tau)$ and hence, by Theorem 15·1, both $L_\tau\{\mathfrak{J}(t)\}$ and $B(\tau)$ are continuous for $\tau > 0$.

We now want to find conditions for $B(\tau)$ which are sufficient to ensure the convergence of the N-series.

19·2. *Preliminary lemmas.*

LEMMA 19·1. *Let* $B(\tau) \geqslant 0$ *in* (16·1). *Then*

(i) *a non-negative solution of* (16·1) *is always the sum of the N-solution and a non-negative solution of the homogeneous equation;*

(ii) *a solution which is* $C(1, 1)$ *is the N-solution.*

If $\mathfrak{J}(\tau)$ is any solution of (16·1), we have by iteration

$$\mathfrak{J}(\tau) = \sum_{\nu=0}^{n} L_\tau^\nu\{B(t)\} + L_\tau^{n+1}\{\mathfrak{J}(t)\}. \tag{19·5}$$

(i) If $\mathfrak{J}(\tau) \geqslant 0$, then

$$\mathfrak{J}(\tau) \geqslant \sum_{\nu=0}^{n} L_\tau^\nu\{B(t)\} = \mathfrak{J}_N^n(\tau).$$

Since $B(\tau) \geqslant 0$, the sequence $\mathfrak{J}_N^n(\tau)$ is monotonic for each fixed τ and converges to $\mathfrak{J}_N(\tau)$ such that $\mathfrak{J}_N(\tau) \leqslant \mathfrak{J}(\tau)$. By Theorem 19·1, $\mathfrak{J}_N(\tau)$ is the N-solution and hence $\mathfrak{J}(\tau) - \mathfrak{J}_N(\tau)$ is a non-negative solution of the homogeneous equation.

(ii) If $\mathfrak{J}(\tau) \in C(1,1)$, then (by Theorem 15·1) $L_\tau\{\mathfrak{J}(t)\} \in C(1,1)$ and therefore $B(\tau) \in C(1,1)$. Let $|\mathfrak{J}(\tau)| \leqslant M$; then by (15·6)

$$|L_\tau^{n+1}\{\mathfrak{J}(t)\}| \leqslant M(2\psi_0)^{n+1} \leqslant M.$$

Hence, from (19·5),

$$0 \leqslant \mathfrak{J}_N^n(\tau) = \sum_{\nu=0}^{n} L_\tau^\nu\{B(t)\} \leqslant 2M.$$

Thus the N-series converges to the N-solution (Theorem 19·1), $\mathfrak{J}_N(\tau)$ is bounded, and therefore (by Theorem 19·2) $\mathfrak{J}_N(\tau) \in C(1,1)$. Hence $\mathfrak{J}(\tau) - \mathfrak{J}_N(\tau)$ is a solution of the homogeneous equation which is $C(1,1)$. By Theorem 18·2, $\mathfrak{J}(\tau) - \mathfrak{J}_N(\tau) = 0$ if $\psi_0 < \frac{1}{2}$. When $\psi_0 = \frac{1}{2}$, by the corollary to Theorem 18·1,

$$\mathfrak{J}(\tau) - \mathfrak{J}_N(\tau) = cf(\tau),$$

and $c = 0$ because $f(\tau) = \tau + O(1)$ for large τ.

Lemma 19·2. *When $\psi_0 = \frac{1}{2}$, the function $q(\tau)$ of Theorem 17·1 is the N-solution of the equation*

$$(1-L)_\tau\{\mathfrak{J}(t)\} = K_3(\tau). \tag{19·6}$$

Since $f(\tau) = \tau + q(\tau)$ and $f(\tau)$ is a solution of (17·4), a simple calculation shows that $q(\tau)$ is a solution of (19·6). Since $q(\tau) \in C(1,1)$, it is the N-solution.

A similar argument proves

Lemma 19·3. *When $\psi_0 = \frac{1}{2}$, the N-solution of*

$$(1-L)_\tau\{\mathfrak{J}(t)\} = K_2(\tau) \tag{19·7}$$

is 1.

If the derivative of $f(\tau)$ exists for $\tau > 0$ and satisfies the conditions of Theorem 15·3, it will follow from (17·4) and the differentiation formula (15·17) that $f'(\tau)/f(0)$ is a solution of

$$(1-L)_\tau\{\mathfrak{J}(t)\} = K_1(\tau). \tag{19·8}$$

It is, in fact, the N-solution, but the proof of the existence of $f'(\tau)$ by the methods of this chapter is long (see Hopf [1], p. 43), whereas it is a simple deduction from the definitive form found for $f(\tau)$ in the next chapter. The consideration of (19·8) will therefore be deferred.

19·3. *An important auxiliary theorem.* Let

$$\mathfrak{J}_r(\tau) = L_\tau\{\mathfrak{J}_r(t)\} + B_r(\tau) \quad (r = 1, 2). \tag{19·9}$$

Then formally (using the notation of § 15·5)

$$(\mathfrak{J}_1, \mathfrak{J}_2) = (\mathfrak{J}_2, L_\tau\{\mathfrak{J}_1\}) + (\mathfrak{J}_2, B_1) = (\mathfrak{J}_1, L_\tau\{\mathfrak{J}_2\}) + (\mathfrak{J}_1, B_2).$$

On using the symmetrical property of L, we have

$$(\mathfrak{J}_2, B_1) = (\mathfrak{J}_1, B_2). \tag{19·10}$$

This formal proof could be substantiated if $\mathfrak{J}_1$ and $\mathfrak{J}_2$ were of integrable square over $(0, \infty)$, but even in the case of equations (19·6) and (19·7), the integral $(\mathfrak{J}_1, \mathfrak{J}_2)$ is divergent. Sufficient conditions for the truth of (19·10) are given by

THEOREM 19·3. *If* $B_r(\tau) \geqslant 0$ $(r = 1, 2)$ *and if* $\mathfrak{J}_r(\tau)$ $(r = 1, 2)$ *are the N-solutions of the equations* (19·9), *then* (19·10) *is true in the sense that, either both integrals are finite and have the same value, or both are infinite.*

Since the operator L is symmetrical in t and τ, so are the operators L^ν $(\nu \geqslant 1)$. By hypothesis

$$\mathfrak{J}_r(\tau) = \sum_{\nu=0}^{\infty} L_\tau^\nu\{B_r(t)\} \quad (r = 1, 2),$$

and therefore

$$(\mathfrak{J}_1, B_2) = \sum_{\nu=0}^{\infty} (B_2, L_\tau^\nu\{B_1\}) = \sum_{\nu=0}^{\infty} (B_1, L_\tau^\nu\{B_2\}) = (\mathfrak{J}_2, B_1)$$

in the stated sense, because all the functions involved are non-negative.

This theorem is true when L is replaced by $\bar{L}$ and the upper limits of integrals by τ_1.

19·4. *The existence of the N-solution* ($\psi_0 = \frac{1}{2}$). We begin with the case $B(\tau) \geqslant 0$. By Lemma 19·1 (i), the existence of the N-solution follows from

THEOREM 19·4. *When* $\psi_0 = \frac{1}{2}$ and $B(\tau) \geqslant 0$, *a necessary and sufficient condition for the equation*

$$(1-L)_\tau\{\mathfrak{J}(t)\} = B(\tau) \tag{19·11}$$

to have a non-negative solution is

$$(B, 1) = \int_0^\infty B(\tau)\, d\tau < \infty. \tag{19·12}$$

Necessity.† If (19·11) has a non-negative solution, by Lemma 19·1 (i) the N-solution $\mathfrak{J}_N(\tau)$ exists. Hence, applying Theorem 19·3 to the equations (19·7) and (19·11), we have

$$\int_0^\infty \mathfrak{J}_N(t)\, K_2(t)\, dt = \int_0^\infty B(t)\, dt. \tag{19·13}$$

Thus we have to show that the integral on the left is finite.

Let $\tau > 0$ (τ fixed) and let $T > \tau$. For $0 \leqslant t \leqslant T$, there is a positive constant A such that $K_1(|t-\tau|) \geqslant A$. Since $K_2(t) \leqslant \psi_0 < 1$ for all t,

$$\begin{aligned}\int_0^T \mathfrak{J}_N(t)\, K_2(t)\, dt &< A^{-1} \int_0^T \mathfrak{J}_N(t)\, K_1(|t-\tau|)\, dt \\ &\leqslant A^{-1}\{\mathfrak{J}_N(\tau) - B(\tau)\}\end{aligned} \tag{19·14}$$

by (19·11) with $\mathfrak{J}(\tau) = \mathfrak{J}_N(\tau)$. For large $t > T$

$$K_1(t-\tau) \sim \Psi(1)\, t^{-1} \exp(\tau - t), \quad K_2(t) \sim \Psi(1)\, t^{-1} \exp(-t),$$

and hence the convergence of

$$\int_T^\infty \mathfrak{J}_N(t)\, K_1(t-\tau)\, dt$$

implies the convergence of

$$\int_T^\infty \mathfrak{J}_N(t)\, K_2(t)\, dt.$$

This, together with (19·14), proves that the left-hand side of (19·13) is finite.

† Proof suggested by E. C. Titchmarsh. The proof of the sufficiency is that given by Hopf in [1].

Sufficiency. Let

$$\mathfrak{J}_N^n(\tau) = \sum_{\nu=0}^{n} L_\tau^\nu\{B(t)\}. \tag{19·15}$$

Because of the symmetry of L^ν,

$$(\mathfrak{J}_N^n, K_2) = \sum_{\nu=0}^{n} (K_2, L_\tau^\nu\{B\}) = \sum_{\nu=0}^{n} (B, L_\tau^\nu\{K_2\})$$

$$= \left(B, \sum_{\nu=0}^{n} L_\tau^\nu\{K_2\}\right) \leqslant (B, 1)$$

by Lemma 19·3. Thus if (19·12) is true,

$$\int_0^\infty \mathfrak{J}_N^n(t)\, K_2(t)\, dt < \infty \tag{19·16}$$

for all n. Since $B(\tau) \geqslant 0$, the sequence $\mathfrak{J}_N^n(\tau)$ is monotonic increasing for each fixed τ, and if it tended to infinity in a set of positive measure, (19·16) would be false. Thus $\mathfrak{J}_N(\tau)$ (the sum of the N-series) exists almost everywhere, and it is the N-solution by Theorem 19·1. Since $\mathfrak{J}_N(\tau)$ is non-negative, the theorem is proved.

From this we can deduce

COROLLARY. *If $\psi_0 = \frac{1}{2}$ and if $B(\tau)$ is $L(0, \infty)$,*† *then the N-series of the equation* (16·1) *converges absolutely for almost all $\tau > 0$ to the N-solution of* (16·1).

For $B(\tau) = B_1(\tau) - B_2(\tau)$, where each $B_r(\tau)$ is non-negative and integrable over $(0, \infty)$. On using the linearity of L, the corollary follows from the theorem.

19·5. *The existence of the N-solution* $(\psi_0 < \frac{1}{2})$. There is no known criterion which is necessary and sufficient for the existence of the N-solution when $\psi_0 < \frac{1}{2}$. The following theorems give sufficient conditions:

THEOREM 19·5. *If $\psi_0 < \frac{1}{2}$ and if $B(\tau) \in C(1, 1)$, then the N-solution of* (16·1) *exists and is $C(1, 1)$.*

Let $|B(\tau)| \leqslant M$; then by (15·6)

$$|L_\tau^\nu\{B(t)\}| \leqslant M(2\psi_0)^\nu,$$

† We shall use the usual notation for Lebesgue integration. There is no fear of confusing this with the operator L.

and hence the N-series converges to a bounded function $\mathfrak{J}_N(\tau)$. If $B(\tau) \geqslant 0$, it follows from Theorems 19·1 and 19·2 that $\mathfrak{J}_N(\tau)$ is the N-solution and that it is $C(1,1)$. If $B(\tau)$ can have either sign, $B(\tau) = B_1(\tau) - B_2(\tau)$, where each $B_r(\tau) \geqslant 0$ and is $C(1,1)$, and the theorem follows on using the linearity of L.

THEOREM 19·6. *If $\psi_0 < \frac{1}{2}$ and if $B(\tau) \in C(1, \tau^p)$, where p is a positive integer, then the N-solution of* (16·1) *exists and is $C(1, \tau^p)$.*

As in Theorem 19·5, it is sufficient to prove the theorem when $B(\tau) \geqslant 0$. Since $B(\tau) \in C(1, \tau^p)$, constants A and B exist such that

$$0 \leqslant B(\tau) \leqslant A + B\tau^p, \tag{19·17}$$

and hence, by (15·6),

$$0 \leqslant L_\tau^\nu\{B(t)\} \leqslant A(2\psi_0)^\nu + BL_\tau^\nu\{t^p\}. \tag{19·18}$$

Thus the N-series will converge to the N-solution and this will be $C(1, \tau^p)$ provided that $\sum_{\nu=0}^{\infty} L_\tau^\nu\{t^p\}$ converges to a function of this class. The proof of this is by induction. It is true when $p = 0$ by Theorem 19·5.

Assume that $\sum_{\nu=0}^{\infty} L_\tau^\nu\{t^p\}$ converges to a function of $C(1, \tau^p)$ when $p \leqslant k-1$ $(k \geqslant 1)$. By the differentiation formula (15·17),

$$\frac{d}{d\tau} L_\tau\{t^k\} = kL_\tau\{t^{k-1}\} \quad (k \geqslant 1). \tag{19·19}$$

Also
$$\begin{aligned} L_0\{t^k\} &= \int_0^\infty t^k K_1(t)\,dt \\ &= \int_0^\infty t^k\,dt \int_1^\infty \Psi(x^{-1})\,x^{-1} \exp(-xt)\,dx \\ &= \int_1^\infty \Psi(x^{-1})\,x^{-1}\,dx \int_0^\infty t^k \exp(-xt)\,dt \\ &= k! \int_0^1 \Psi(u)\,u^k\,du \\ &= k!\,\psi_k. \end{aligned} \tag{19·20}$$

Hence
$$L_\tau\{t^k\} = k!\,\psi_k + k\int_0^\tau L_u\{t^{k-1}\}\,du. \tag{19·21}$$

Since
$$L_\tau\{1\} = 2\psi_0 - K_2(\tau),$$
therefore, by (19·21),
$$L_\tau\{t^k\} = 2\psi_0 \tau^k + a_1 \tau^{k-1} + \ldots + a_k + (-1)^{k-1} k!\, K_{k+2}(\tau), \quad (19\cdot 22)$$
where $a_1, \ldots, a_k$ are constants. It follows that
$$L_\tau^{\nu+1}\{t^k\} - 2\psi_0 L_\tau^\nu\{t^k\}$$
$$= a_1 L_\tau^\nu\{t^{k-1}\} + \ldots + a_k L_\tau^\nu\{1\} + (-1)^{k-1} k!\, L_\tau^\nu\{K_{k+2}(t)\}.$$
Summing for $\nu = 0$ to n, we get
$$(1-2\psi_0)\sum_{\nu=0}^{n} L_\tau^\nu\{t^k\} + L_\tau^{n+1}\{t^k\} = \tau^k + a_1 \sum_{\nu=0}^{n} L_\tau^\nu\{t^{k-1}\} + \ldots$$
$$+ a_k \sum_{\nu=0}^{n} L_\tau^\nu\{1\} + (-1)^{k-1} k! \sum_{\nu=0}^{n} L_\tau^\nu\{K_{k+2}(t)\}. \quad (19\cdot 23)$$
Since $K_{k+2}(\tau) \in C(1,1)$, the last series on the right converges, as $n \to \infty$, to a function of $C(1,1)$ and hence, using the hypothesis, the right-hand side of (19·23) converges to a function of $C(1,\tau^k)$. Since $L_\tau^{n+1}\{t^k\} > 0$ and $(1-2\psi_0) > 0$, $\sum_{\nu=0}^{n} L_\tau^\nu\{t^k\}$ must converge, as $n \to \infty$, to a function of $C(1,\tau^k)$.

It now follows by induction that $\sum_{\nu=0}^{\infty} L_\tau^\nu\{t^p\}$ converges to a function of $C(1,\tau^p)$ for $p = 1, 2, \ldots$, and this completes the proof of the theorem.

Chapter 5

THE WIENER–HOPF SOLUTION

20. Introduction

The Wiener–Hopf method was designed by its authors as a means of obtaining an explicit formula for the solution of homogeneous equations such as the Milne equation

$$J(\tau) = \int_0^\infty J(t)\, K_1(|t-\tau|)\, dt. \qquad (20\cdot1)$$

(We use $J(\tau)$ to denote any solution of this equation.) The method has been used for investigating the solutions of equations whose kernels are subject to less restrictive conditions than $K_1(\tau)$ (see, for example, Smithies [1], where a long list of early references is given).

By various ingenious devices, atomic physicists have used the Wiener–Hopf method as a means of solving particular cases of the non-homogeneous equation (see Davison [1], Elliott [1]), but it is doubtful whether the method is then as good as that developed in Chapter 6.

The Wiener–Hopf method will be explained here in the simple case of (20·1). In the form given by its authors (see Wiener and Hopf [1]), the solution was based on the two-sided Laplace transform. In order to solve (20·1), it would be written in the form

$$F(\tau) = J(\tau) - \int_{-\infty}^\infty J(t)\, K_1(|t-\tau|)\, dt, \qquad (20\cdot2)$$

where $J(\tau) = 0$ for $\tau < 0$ and $F(\tau) = 0$ for $\tau > 0$. Relations would then be obtained between the two-sided transforms of $F(\tau)$, $J(\tau)$ and $K_1(|\tau|)$.

A modified version of the method was given by Titchmarsh in [2], § 11·17, and it is this form which has been widely used by atomic physicists. Titchmarsh employs Fourier transforms of the form

$$\left.\begin{aligned} F^+(\tau) &= \frac{1}{\sqrt{(2\pi)}}\int_0^\infty f(x)\exp(ix\tau)\,dx, \\ F^-(\tau) &= \frac{1}{\sqrt{(2\pi)}}\int_{-\infty}^0 f(x)\exp(ix\tau)\,dx, \end{aligned}\right\} \quad (20\cdot3)$$

where τ is a complex variable. Since, however, Laplace transforms arise naturally in transfer theory, the most satisfactory treatment seems to be to follow Titchmarsh's method, but with the functions (replacing f by J)

$$\left.\begin{aligned} j^+(s) &= s\int_0^\infty J(t)\exp(-st)\,dt, \\ j^-(s) &= s\int_{-\infty}^0 J(t)\exp(-st)\,dt \end{aligned}\right\} \quad (20\cdot4)$$

in place of (20·3).

21. The Laplace transforms (20·4)

The following results are easily deduced from Widder [1], chap. II, § 10:

If $J(t)\exp(-ct)$ is $L^2(0,\infty)$, the integral for $s^{-1}j^+(s)$ converges in mean on the line $\operatorname{re} s = c$ and absolutely for $\operatorname{re} s > c$. The function $s^{-1}j^+(s)$ is $L^2(-\infty,\infty)$ on $\operatorname{re} s = c$ and is regular for $\operatorname{re} s > c$. Similarly, if $J(t)\exp(-c't)$ is $L^2(-\infty, 0)$, $s^{-1}j^-(s)$ exists as a limit in mean on $\operatorname{re} s = c'$ and is regular for $\operatorname{re} s < c'$.

Subject to these conditions, the formulae (20·4) can be inverted to give (almost everywhere)

$$\underset{\omega\to\infty}{\text{l.i.m.}}\frac{1}{2\pi i}\int_{c-i\omega}^{c+i\omega} s^{-1}j^+(s)\exp(s\tau)\,ds = \begin{cases} J(\tau) & (\tau>0), \\ 0 & (\tau<0), \end{cases} \quad (21\cdot1)$$

$$\underset{\omega\to\infty}{\text{l.i.m.}}\frac{1}{2\pi i}\int_{c'-i\omega}^{c'+i\omega} s^{-1}j^-(s)\exp(s\tau)\,ds = \begin{cases} 0 & (\tau>0), \\ J(\tau) & (\tau<0). \end{cases} \quad (21\cdot2)$$

Conversely, if $s^{-1}j^+(s)$ is $L^2(-\infty,\infty)$ on $\operatorname{re} s = c$, and if $J(\tau)$ is defined, for $\tau > 0$, by (21·1), then $J(\tau)\exp(-c\tau)$ is $L^2(0,\infty)$. With the notation of (5·16), we shall write

$$j^+(s) = \mathfrak{L}_s\{J(\tau)\}, \quad J(\tau) = \mathfrak{L}_\tau^{-1}\{j^+(s)\}.$$

If, for some c, $J_1(t)\exp(-ct)$ and $J_2(t)\exp(ct)$ are $L^2(-\infty,\infty)$, and if $j_r^+(s)$, $j_r^-(s)$ $(r=1,2)$ are the corresponding Laplace transforms for the appropriate values of s, then the Parseval theorem is true in the form

$$\int_{-\infty}^{\infty} J_1(t)\,J_2(t)\,dt$$
$$= -\frac{1}{2\pi i}\int_{c-i\infty}^{c+i\infty}\{j_1^+(s)+j_1^-(s)\}\{j_2^+(-s)+j_2^-(-s)\}\frac{ds}{s^2}, \quad (21\cdot3)$$

both integrals being absolutely convergent. In particular, if $J_1(\tau)=0$ for $\tau<0$, then

$$\int_0^{\infty} J_1(t)\,J_2(t)\,dt = -\frac{1}{2\pi i}\int_{c-i\infty}^{c+i\infty} j_1^+(s)\{j_2^+(-s)+j_2^-(-s)\}\frac{ds}{s^2}. \quad (21\cdot4)$$

22. The homogeneous Milne equation

Working formally for the moment and assuming (20·1) is true for all real τ, if we multiply by $s\exp(-s\tau)$, integrate over $(-\infty,\infty)$ and invert the order of integration, we obtain

$$j^+(s)+j^-(s) = s\int_0^{\infty} J(t)\exp(-st)\,dt$$
$$\times\int_{-\infty}^{\infty} K_1(|t-\tau|)\exp[-s(\tau-t)]\,d\tau$$
$$= j^+(s)\int_{-\infty}^{\infty} K_1(|x|)\exp(-sx)\,dx. \quad (22\cdot1)$$

If $-1<\operatorname{re} s<1$,

$$\int_{-\infty}^{\infty} K_1(|x|)\exp(-sx)\,dx$$
$$= \int_{-\infty}^{\infty}\exp(-sx)\,dx\int_1^{\infty}\Psi(t^{-1})\exp(-|x|\,t)\frac{dt}{t}$$
$$= \int_1^{\infty}\Psi(t^{-1})\frac{dt}{t}\left\{\int_0^{\infty}\exp[-(s+t)x]\,dx+\int_0^{\infty}\exp[-(t-s)x]\,dx\right\}$$
$$= \int_1^{\infty}\Psi(t^{-1})\left\{\frac{1}{s+t}+\frac{1}{t-s}\right\}\frac{dt}{t}$$
$$= 2\int_0^1\frac{\Psi(y)}{1-s^2y^2}\,dy$$
$$= 1-T(s^{-1}), \quad (22\cdot2)$$

where $T(\mu)$ is defined by (8·4). Hence (22·1) gives

$$j^-(s) = -j^+(s)\,T(s^{-1}). \qquad (22\cdot3)$$

Thus the solution of (20·1) will depend on the function $T(\mu)$ studied in § 9. Also $s^{-1}j^+(s)$ is regular in some half-plane $\operatorname{re} s > c$, and $s^{-1}j^-(s)$ is regular in some half-plane $\operatorname{re} s < c'$. Hence, if (20·1) has a non-zero solution, it must be possible to split $T(s^{-1})$ into the quotient of two such functions. This was, in fact, done in § 11, and some of the work which follows will repeat, with fewer details, a part of what was done there. Since § 11 led to the H-functions, $j^+(s)$ and $j^-(s)$ will be expressed in terms of $H(s^{-1})$. When $j^+(s)$ has been found, $J(\tau)$ for $\tau > 0$ will be given by the inversion formula (21·1).

22·1. *The existence theorems.* We shall prove the following theorems:

THEOREM 22·1. *Let* $\psi_0 = \frac{1}{2}$; *then the equation* (20·1) *has a non-zero solution* $J(\tau)$ *in the class* $C(\ln \tau^{-1}, \tau^\alpha)$ *provided that* $\alpha \geqslant 1$. *If* $0 \leqslant \alpha < 1$, *then* $J(\tau) = 0$.

THEOREM 22·2. *Let* $0 < \psi_0 < \frac{1}{2}$; *then the equation* (20·1) *has a non-zero solution* $J(\tau)$ *in the class* $C(\ln \tau^{-1}, \exp a\tau)$ *provided that* $k \leqslant a < 1$, *where* $\mu = \pm k^{-1}$ *are the roots of* $T(\mu) = 0$. *If* $0 \leqslant a < k$, *then* $J(\tau) = 0$.

Since the class $C(\ln \tau^{-1}, \tau^\alpha)$ is a sub-class of $C(\ln \tau^{-1}, \exp a\tau)$ $(0 < a < 1)$, except where indicated, we shall work with the wider class and the analysis will apply to both cases.

Let $J(\tau)$ be defined for $\tau < 0$ by (20·1). Writing $-\tau$ for τ, we have

$$J(-\tau) = \int_0^\infty J(t)\,K_1(t+\tau)\,dt \quad (\tau > 0). \qquad (22\cdot4)$$

For large τ, by (14·10),

$$|J(-\tau)| = O\left[\exp(-\tau)\int_0^\infty |J(t)|\exp(-t)\,dt\right]$$

$$= O[\exp(-\tau)], \qquad (22\cdot5)$$

since $J(t) \in C(\ln t^{-1}, \exp at)$ and $a < 1$. For small τ

$$|J(-\tau)| = O\left(\int_0^{\frac{1}{2}} |J(t)| \ln(t+\tau)^{-1} dt\right) + O\left(\int_{\frac{1}{2}}^{\infty} |J(t)| \exp(-t)\, dt\right)$$

$$= O\left(\ln \tau^{-1} \int_0^{\frac{1}{2}} |J(t)|\, dt\right) + O(1)$$

$$= O(\ln \tau^{-1}). \tag{22·6}$$

From (22·5) and (22·6) it follows that $J(\tau)\exp(-c'\tau)$ is $L^2(-\infty, 0)$ for every $c' < 1$. Thus $s^{-1}j^-(s)$ exists and is regular for $\operatorname{re} s < 1$. Since $J(\tau)\exp(-c\tau)$ is $L^2(0,\infty)$ for every $c > a$, $s^{-1}j^+(s)$ exists and is regular for $\operatorname{re} s > a$. These two half-planes have the strip $a < \operatorname{re} s < 1$ in common, and in this strip the inversion of the order of integration in (22·1) is justified by absolute convergence. Hence the equation (22·3) is true for $a < \operatorname{re} s < 1$. Thus we now want to express $T(s^{-1})$ as the quotient of a function regular for $\operatorname{re} s < 1$ by one regular for $\operatorname{re} s > a$.

Let†

$$\Phi(s) = \frac{s^2 - 1}{s^2 - k^2} T(s^{-1}), \tag{22·7}$$

where $k = 0$ when $\psi_0 = \frac{1}{2}$, and $\pm k$ $(0 < k < 1)$ are the zeros of $T(s^{-1})$ when $\psi_0 < \frac{1}{2}$. By § 9·1, $\Phi(s)$ is regular and not zero in the s-plane cut along $(-\infty, -1)$ and $(1, \infty)$, and

$$\Phi(s) = 1 + O(|s|^{-1}) \quad \text{as} \quad |s| \to \infty \tag{22·8}$$

uniformly in $\arg s$ in the cut plane. Since $\Phi(s)$ is real and positive on the imaginary axis, there is no variation in $\arg \Phi(s)$ along it and $\log \Phi(s)$ can therefore be defined to be such that

$$\log \Phi(s) = O(|s|^{-1}) \quad \text{as} \quad |s| \to \infty \tag{22·9}$$

uniformly in $\arg s$. This branch is regular in the cut plane.

Let $|\operatorname{re} s| < b$, where $0 < b < 1$. In view of (22·9), Cauchy's integral theorem gives

$$\log \Phi(s) = \frac{1}{2\pi i}\left\{\int_{b-i\infty}^{b+i\infty} - \int_{-b-i\infty}^{-b+i\infty}\right\} \frac{\log \Phi(z)}{z - s}\, dz,$$

† Cf. (11·3). Since, by the hypothesis (14·1), $\Psi(1) > 0$, case (iii) of Chap. 2 does not arise. (See the end of § 9.)

where the z-plane is cut similarly. Defining $\Phi^+(s)$ and $\Phi^-(s)$ by

$$\log \Phi^+(s) = \frac{1}{2\pi i}\int_{-b-i\infty}^{-b+i\infty} \frac{\log \Phi(z)}{z-s}\,dz, \tag{22·10}$$

$$\log \Phi^-(s) = \frac{1}{2\pi i}\int_{b-i\infty}^{b+i\infty} \frac{\log \Phi(z)}{z-s}\,dz, \tag{22·11}$$

we have
$$\Phi(s) = \frac{\Phi^-(s)}{\Phi^+(s)}, \tag{22·12}$$

and this, with (22·7), gives the required decomposition of $T(s^{-1})$. From (22·10), $\log \Phi^+(s)$ is regular for $\operatorname{re} s > -b$. Since (by Cauchy's theorem) b can be taken as near to -1 as we please, $\log \Phi^+(s)$ is regular for $\operatorname{re} s > -1$, and $\Phi^+(s)$ is regular and not zero for $\operatorname{re} s > -1$. Similarly, $\Phi^-(s)$ is regular and not zero for $\operatorname{re} s < 1$.

By (22·7) and (22·12), equation (22·3) can be written

$$\frac{s^{-1}j^+(s)}{\Phi^+(s)} \cdot \frac{s^2-k^2}{s+1} = -\frac{s^{-1}j^-(s)}{\Phi^-(s)}(s-1). \tag{22·13}$$

Since $\Phi^+(s)$ and $\Phi^-(s)$ are non-zero in their respective half-planes, the left-hand side of (22·13) is regular for $\operatorname{re} s > a$ and the right-hand side for $\operatorname{re} s < 1$. Since there is a strip $a < \operatorname{re} s < 1$ in common, each side gives the analytic continuation of the other and each is equal to an integral function $\chi(s)$. We shall first show that $\chi(s) = As + B$, where A and B are constants, and then that $A = 0$.

Let $\sigma = \operatorname{re} s \geqslant \frac{1}{2}(a+1)$. Since, for $\tau > 0$, $J(\tau) \in C(\ln \tau^{-1}, \exp a\tau)$, therefore

$$\begin{aligned} |s^{-1}j^+(s)| &\leqslant \int_0^\infty |J(t)| \exp(-\sigma t)\,dt \\ &= O\left(\int_0^\eta \ln t^{-1}\,dt\right) + O\left(\int_\eta^\infty \exp[-(\sigma-a)t]\,dt\right) \\ &= o(1) + O[(\sigma-a)^{-1}] \end{aligned}$$

by choosing η small enough. Hence

$$|s^{-1}j^+(s)| = O(1) \quad [\operatorname{re} s \geqslant \tfrac{1}{2}(a+1)], \tag{22·14}$$

and
$$s^{-1}j^+(s) \to 0 \quad \text{as} \quad \operatorname{re} s \to \infty. \tag{22·15}$$

By (22·9), $\log\Phi(-b+iy)$ is $L^2(-\infty,\infty)$ and hence, by (22·10) and Schwarz's inequality,

$$|\log\Phi^+(s)| \leqslant \frac{1}{2\pi}\left(\int_{-\infty}^{\infty}|\log\Phi(-b+iy)|^2\,dy\right)^{\frac{1}{2}}$$

$$\times\left(\int_{-\infty}^{\infty}\frac{dy}{(b+\sigma)^2+(y-t)^2}\right)^{\frac{1}{2}}$$

$$= O[(b+\sigma)^{-\frac{1}{2}}] \quad (s=\sigma+it). \tag{22·16}$$

Since b may be as near to 1 as we please, $\log\Phi^+(s)$ is bounded for $\mathrm{re}\,s \geqslant -1+\delta$ $(\delta>0)$ and therefore both $\Phi^+(s)$ and $1/\Phi^+(s)$ are bounded. Moreover, $\log\Phi^+(s)\to 0$ as $\sigma\to\infty$ and so

$$\Phi^+(s)\to 1 \quad \text{as} \quad \mathrm{re}\,s\to\infty. \tag{22·17}$$

By considering the left-hand side of (22·13), it is seen that $s^{-1}\chi(s)$ is bounded as $|s|\to\infty$ for $\mathrm{re}\,s\geqslant\frac{1}{2}(a+1)$ and that

$$s^{-1}\chi(s)\to 0 \quad \text{as} \quad \mathrm{re}\,s\to\infty. \tag{22·18}$$

Similar proofs, using (22·5) and (22·6) in place of the order conditions for $\tau>0$, show that $s^{-1}j^-(s)$ is bounded for $\mathrm{re}\,s\leqslant\frac{1}{2}(a+1)$ and that both $\Phi^-(s)$ and $1/\Phi^-(s)$ are bounded for $\mathrm{re}\,s\leqslant 1-\delta$ $(\delta>0)$. Hence $s^{-1}\chi(s)$ is also bounded as $|s|\to\infty$ for $\mathrm{re}\,s\leqslant\frac{1}{2}(a+1)$. It follows that $s^{-1}\chi(s)$ is bounded as $s\to\infty$ in any manner and hence that $\chi(s)=As+B$. However, for (22·18) to be true, we must have $A=0$ and hence each side of (22·13) is a constant B. In particular

$$j^+(s)=\frac{Bs(s+1)}{s^2-k^2}\Phi^+(s). \tag{22·19}$$

It is now necessary to consider the cases $\psi_0=\frac{1}{2}$ and $\psi_0<\frac{1}{2}$ separately.

Case (i). $\psi_0=\frac{1}{2}$. Then $k=0$ in (22·19) and

$$j^+(s)=Bs^{-1}(s+1)\Phi^+(s). \tag{22·20}$$

Since $\Phi^+(0)$ is finite and not zero, $sj^+(s)$ tends to a finite non-zero limit as $s\to 0$ unless $B=0$. But, if $0<s<1$, we obtain from (20·4) and the order conditions of Theorem 22·1

$$|s^\alpha j^+(s)| = s^{\alpha+1}\left\{O\left(\int_0^1 \ln \tau^{-1}\, d\tau\right) + O\left(\int_1^\infty \tau^\alpha \exp(-s\tau)\, d\tau\right)\right\}$$
$$= s^{\alpha+1}\{O(1)+O(s^{-\alpha-1})\}$$
$$= O(1).$$

Hence, if $\alpha < 1$,

$$|sj^+(s)| = O(s^{1-\alpha}) \to 0 \quad \text{as} \quad s \to 0,$$

and therefore $B = 0$. Thus $j^+(s) = 0$ and $J(\tau) = 0$ when $\alpha < 1$.

Case (ii). $\psi_0 < \frac{1}{2}$. If $a < k$, then (by § 21) $j^+(s)$ is regular at $s = k$. Since $\Phi^+(s)$ is regular and not zero at $s = k$, the right-hand side of (22·19) has a pole there unless $B = 0$. Hence $j^+(s) = 0$ and $J(\tau) = 0$.

When $\alpha \geqslant 1$ in case (i), or when $k \leqslant a < 1$ in case (ii), there is no contradiction and it remains to verify that the function $J(\tau) = \mathfrak{L}_\tau^{-1}\{j^+(s)\}$, where $j^+(s)$ is given by (22·19), does satisfy (20·1). Later (§ 23) the assumed order conditions will be verified.

In Parseval's theorem (21·4) take

$$J_1(t) = J(t), \quad J_2(t) = K_1(|t-\tau|),$$

where $\tau > 0$. Let $k < c < 1$, where $k = 0$ when $\psi_0 = \frac{1}{2}$. Since $\Phi^+(s)$ is bounded on the line $\operatorname{re} s = c$, $s^{-1}j^+(s)$ is $L^2(-\infty, \infty)$ and therefore $J(t)\exp(-ct)$ is $L^2(0,\infty)$. From the order conditions for $K_1(t)$ it follows that

$$K_1(|t-\tau|)\exp(ct) \quad \text{is} \quad L^2(-\infty,\infty);$$

also (with the notation of § 21), if $-1 < \operatorname{re} s < 1$,

$$j_2^+(s) + j_2^-(s) = s\int_{-\infty}^{\infty} K_1(|t-\tau|)\exp(-st)\, dt$$
$$= s[1 - T(s^{-1})]\exp(-s\tau)$$

by (22·2). Hence Parseval's theorem gives

$$\int_0^\infty J(t)\, K_1(|t-\tau|)\, dt = \frac{1}{2\pi i}\int_{c-i\infty}^{c+i\infty} j^+(s)\,[1 - T(-s^{-1})]\exp s\tau \frac{ds}{s}$$
$$= J(\tau) - I(\tau), \qquad (22{\cdot}21)$$

where, by (22·19),

$$I(\tau) = \frac{B}{2\pi i}\int_{c-i\infty}^{c+i\infty} \frac{s+1}{s^2-k^2}\Phi^+(s)\,T(-s^{-1})\exp(s\tau)\,ds, \qquad (22\cdot22)$$

and the integrals for $J(\tau)$ and $I(\tau)$ may only exist as limits in mean. Since $T(-s^{-1}) = T(s^{-1})$, it follows from (22·7) and (22·12) that

$$\Phi^+(s)\,T(-s^{-1})\,(s+1)/(s^2-k^2) = \Phi^-(s)/(s-1). \qquad (22\cdot23)$$

Since $\Phi^-(s)$ is bounded for $\operatorname{re} s \leqslant c$,

$$|\Phi^-(s)/(s-1)| = O(|s|^{-1}) \quad (\operatorname{re} s \leqslant c) \qquad (22\cdot24)$$

for large $|s|$, and it therefore follows that $I(\tau)$ exists in the form

$$I(\tau) = \lim_{\omega\to\infty} \frac{B}{2\pi i}\int_{c-i\omega}^{c+i\omega} \frac{\Phi^-(s)}{s-1}\exp(s\tau)\,ds, \qquad (22\cdot25)$$

and that $J(\tau)$ also exists as a Cauchy principal value. The function $\Phi^-(s)/(s-1)$ is regular for $\operatorname{re} s \leqslant c$ and satisfies (22·24), and hence (by Cauchy's theorem and a lemma given by Carslaw and Jaeger in [1]†) $I(\tau) = 0$. Thus (22·21) is (20·1).

23. The definitive form for $J(\tau)$

Consider the solution of (20·1) which is such that $J(0) = 1$, where $J(0)$ denotes $\lim_{\tau\to+0} J(\tau)$. Any other solution will be $J(0)$ times this solution. It is well known (subject to continuity conditions which are satisfied here) that, for real s,

$$\lim_{s\to\infty} j^+(s) = \lim_{\tau\to+0} J(\tau),$$

and hence, from (22·17) and (22·19), $B = 1$ and

$$J(\tau) = \lim_{\omega\to\infty} \frac{1}{2\pi i}\int_{c-i\omega}^{c+i\omega} \frac{s+1}{s^2-k^2}\Phi^+(s)\exp(s\tau)\,ds, \qquad (23\cdot1)$$

where $k < c$ ($k = 0$ when $\psi_0 = \frac{1}{2}$). We shall now transform this into a real form with the aid of the H-function.

† Let C_1 and C_2 be the arcs of $|s| = R$ for which $\cos^{-1}(c/R) \leqslant |\arg s| \leqslant \pi$ and let $f(s)$ be an analytic function satisfying the condition $|f(s)| = O(R^{-k})$ ($k > 0$) on C_1 and C_2. Then

$$\int f(s)\exp(st)\,ds$$

taken along C_1 and C_2 tends to zero as $R\to\infty$, provided that $t > 0$.

The functions $\Phi(s)$, $\Phi^+(s)$, $\Phi^-(s)$ of §§ 11 and 22 are the same. Hence, by (11·14),

$$\Phi^+(s) = (s+k)\,H(s^{-1})/(s+1). \tag{23·2}$$

Thus (22·19) is

$$j^+(s) = sH(s^{-1})/(s-k) \tag{23·3}$$

and (23·1) is

$$J(\tau) = \lim_{\omega\to\infty} \frac{1}{2\pi i}\int_{c-i\omega}^{c+i\omega} \frac{H(s^{-1})}{s-k}\exp(s\tau)\,ds. \tag{23·4}$$

The integrand in (23·4) is regular for $\operatorname{re} s \geqslant c$, but it has poles in $\operatorname{re} s < c$. Moreover, the s-plane must be cut along $(-\infty, -1)$. When $\psi_0 < \frac{1}{2}$, $H(s^{-1})$ is regular and not zero at $s = k$ and therefore $s = k$ is a simple pole of the integrand. By (10·1),

$$H(s^{-1})/(s-k) = 1/\{(s-k)\,H(-s^{-1})\,T(-s^{-1})\}. \tag{23·5}$$

$T(-s^{-1})$ has a simple zero at $s = -k$ and therefore $s = -k$ is also a simple pole of the integrand. When $\psi_0 = \frac{1}{2}$, these poles coincide at $s = 0$.

23·1. *The residue at $s = 0$ when $\psi_0 = \frac{1}{2}$.* From (11·37)

$$\begin{aligned} 1/H(s^{-1}) &= s\int_0^1 xH(x)\,\Psi(x)\,(1+sx)^{-1}\,dx \\ &= s[h_1 - sh_2 + \dots] \quad (|s| < 1), \end{aligned} \tag{23·6}$$

where h_n is defined by (12·2). Hence, for sufficiently small $|s|$,

$$H(s^{-1}) = s^{-1}h_1^{-1}[1 + sh_2/h_1 + \dots]. \tag{23·7}$$

Since $k = 0$, the integrand in (23·4) is

$$s^{-1}H(s^{-1})\exp(s\tau) = s^{-2}h_1^{-1}\{1 + sh_2/h_1 + \dots\}\{1 + s\tau + \dots\}$$

and the residue at the pole is therefore

$$\Sigma_0 = h_1^{-1}(\tau + h_2/h_1). \tag{23·8}$$

By (12·5) with $\psi_0 = \frac{1}{2}$ and $n = 1$, $h_1 = \sqrt{(2\psi_2)}$. Hence

$$\Sigma_0 = (2\psi_2)^{-\frac{1}{2}}\,(\tau + \tau_0), \tag{23·9}$$

where

$$\tau_0 = (2\psi_2)^{-\frac{1}{2}}\,h_2. \tag{23·10}$$

In neutron-diffusion theory, τ_0 is known as the extrapolated end-point.

23·2. *The sum of the residues when $\psi_0 < \frac{1}{2}$.* The residue of the integrand in (23·4) at $s = k$ is

$$H(k^{-1}) \exp(k\tau).$$

On rewriting the integrand by means of (23·5), the residue at $s = -k$ is seen to be

$$-\tfrac{1}{2}k \exp(-k\tau)/\{H(k^{-1})\, T'(k^{-1})\},$$

where $T'(\mu) = (d/d\mu)\, T(\mu)$. Thus the sum of the residues is

$$\Sigma_k = c_k \sinh[k(\tau+\tau_k)], \tag{23·11}$$

where

$$c_k = \sqrt{\{2k/T'(k^{-1})\}}, \tag{23·12}$$

$$\tau_k = \tfrac{1}{2}k^{-1} \ln\{2k^{-1}[H(k^{-1})]^2\, T'(k^{-1})\}. \tag{23·13}$$

The constant τ_k is also known as the extrapolated end-point. In each case, τ_k is the distance above the surface $\tau = 0$ at which $\Sigma_k = 0$.

23·3. *The transformation of* (23·4).† Let Γ be the closed contour consisting of:

(i) the line $\operatorname{re} s = c$ from $c - i\omega$ to $c + i\omega$;

(ii) the arc C_1 of $|s| = R = \sqrt{(c^2+\omega^2)}$ for $\cos^{-1} c/R \leqslant \arg s \leqslant \pi$;

(iii) the upper side of the cut along $(-\infty, -1)$ from $s = -R$ to $s = -1-\delta$;

(iv) the circle γ given by $|s+1| = \delta$ from $\arg(s+1) = \pi$ to $\arg(s+1) = -\pi$;

(v) the lower side of the cut from $s = -1-\delta$ to $s = -R$;

(vi) the arc C_2 of $|s| = R$ for $-\pi \leqslant \arg s \leqslant -\cos^{-1} c/R$.

By Cauchy's residue theorem (using (23·5) to transform the integrand)

$$\frac{1}{2\pi i}\int_\Gamma \{(s-k)\, H(-s^{-1})\, T(-s^{-1})\}^{-1} \exp(s\tau)\, ds = \Sigma_k. \tag{23·14}$$

By (9·6) and (11·15)

$$\{(s-k)\, H(-s^{-1})\, T(-s^{-1})\}^{-1} = O(|s|^{-1})$$

on C_1 and C_2 and hence, by Carslaw and Jaeger's lemma (see the footnote on p. 55), the integrals along C_1 and C_2 tend to zero as $R \to \infty$. On γ the integrand is bounded and the integral round γ therefore tends to zero as $\delta \to 0$.

† Cf. Mark [1]

Thus, as $R \to \infty$, $\delta \to 0$, the integral on the left of (23·14) reduces to $J(\tau)$ plus the integrals along the sides of the cut. On making the transformation $s = -1/z$, the cut goes into one along (0, 1) in the z-plane and (23·14) becomes

$$J(\tau) = \Sigma_k - \frac{1}{2\pi i} \int_{\lambda \cup \lambda'} \frac{\exp(-\tau/z)}{z(1+kz)\, H(z)\, T(z)}\, dz, \qquad (23\cdot15)$$

where the integral is taken along λ (the lower side of the cut) from 0 to 1, and along λ' (the upper side of the cut) from 1 to 0. On λ (for details see the proof of (11·26))

$$T(z) = T_c(x) - i\pi x \Psi(x),$$

where $T_c(x)$ is the Cauchy principal value of $T(z)$ at $z = x$ $(0 < x < 1)$; on λ'

$$T(z) = T_c(x) + i\pi x \Psi(x).$$

The other parts of the integrand are one-valued on the cut and hence

$$J(\tau) = \Sigma_k - \int_0^1 \frac{\Psi(x) \exp(-\tau/x)}{(1+kx)\, H(x)\, Z(x)}\, dx, \qquad (23\cdot16)$$

where†

$$Z(x) = [T_c(x)]^2 + \pi^2 x^2 [\Psi(x)]^2. \qquad (23\cdot17)$$

This is the final form for $J(\tau)$. When $\psi_0 = \frac{1}{2}$, $k = 0$, and Σ_0 is given by (23·9); when $\psi_0 < \frac{1}{2}$, Σ_k is given by (23·11).

For large τ, the integral in (23·16) is $O[\exp(-\tau)]$ and hence

$$J(\tau) = \Sigma_k - O[\exp(-\tau)] \quad \text{as} \quad \tau \to \infty. \qquad (23\cdot18)$$

Since $J(0) = 1$, the order conditions assumed in Theorems 22·1 and 22·2 are satisfied.

Collecting all these results, we have

THEOREM 23·1. *Let $\psi_0 = \frac{1}{2}$ and let $J(\tau)$ be the solution of* (20·1) *in the class $C(1, \tau)$ such that $J(0) = 1$. Then $J(\tau)$ is given by* (23·16), *where $k = 0$ and Σ_0 is defined by* (23·9). *If*

$$j(s) = \mathfrak{L}_s\{J(\tau)\},$$

then

$$j(\mu^{-1}) = H(\mu) \quad (\mu \geqslant 0). \qquad (23\cdot19)$$

Any non-null solution of (20·1) *in the class $C(\ln \tau^{-1}, \tau^{\alpha})$ $(\alpha \geqslant 1)$ is a constant multiple of this solution.*

† In general, $T_c(x)$ and $\Psi(x)$ will not vanish together. Since $T_c(0) = 1$, therefore $Z(x) > 0$ for $0 \leqslant x \leqslant 1$.

THEOREM 23·2. *Let $\psi_0 < \frac{1}{2}$ and let k^{-1} be the positive zero of $T(\mu)$, and let $J(\tau)$ be the solution of* (20·1) *in the class $C(1, \exp k\tau)$ such that $J(0) = 1$. Then $J(\tau)$ is given by* (23·16), *where Σ_k is defined by* (23·11). *If $j(s) = \mathfrak{L}_s\{J(\tau)\}$, then*

$$j(\mu^{-1}) = H(\mu)/(1-k\mu) \quad (\mu \geqslant 0). \tag{23·20}$$

Any non-null solution of (20·1) *in the class $C(\ln \tau^{-1}, \exp a\tau)$ $(k \leqslant a < 1)$ is a constant multiple of this solution.*

When $\psi_0 = \frac{1}{2}$, by (23·9) and (23·18),

$$J(\tau) = (2\psi_2)^{-\frac{1}{2}}(\tau+\tau_0) - O[\exp(-\tau)]. \tag{23·21}$$

On comparing this with (17·5), it is seen that

$$J(\tau) = (2\psi_2)^{-\frac{1}{2}} f(\tau) = (2\psi_2)^{-\frac{1}{2}}[\tau + q(\tau)], \tag{23·22}$$

where $q(\tau)$ is Hopf's function. Also, from (23·16),

$$q(\tau) = \tau_0 - (2\psi_2)^{\frac{1}{2}} \int_0^1 \frac{\Psi(x)\exp(-\tau/x)}{H(x)Z(x)}\, dx, \tag{23·23}$$

so that $\tau_0 = q(\infty)$. Finally, since $J(0) = 1$, (23·22) gives

$$f(0) = q(0) = (2\psi_2)^{\frac{1}{2}}. \tag{23·24}$$

24. The derivatives of $J(\tau)$

When τ is complex, the integrand in (23·16) is a continuous function of x and τ for $\operatorname{re}\tau > 0$ and $0 \leqslant x \leqslant 1$, and for each x in $(0, 1)$ it is regular for $\operatorname{re}\tau > 0$. When $\operatorname{re}\tau \geqslant \delta > 0$, the integrand is less than $\Psi(x)/[H(x)Z(x)]$, which is continuous in the interval $0 \leqslant x \leqslant 1$, and the integral therefore converges uniformly. Hence $J(\tau) - \Sigma_k$ is regular for $\operatorname{re}\tau > 0$ and all derivatives can be found by differentiation under the integral sign. Since Σ_k is an integral function, $J(\tau)$ is regular for $\operatorname{re}\tau > 0$ and

$$J'(\tau) = \Sigma_k' + \int_0^1 \frac{\Psi(x)\exp(-\tau/x)}{x(1+kx)H(x)Z(x)}\, dx. \tag{24·1}$$

When τ is real $(\tau > 0)$, the integrand in (24·1) is non-negative, and hence

$$\Sigma_k' \leqslant J'(\tau) \leqslant \Sigma_k' + O\left[\int_0^1 \Psi(x)\, x^{-1} \exp(-\tau/x)\, dx\right],$$

i.e.

$$\Sigma_k' \leqslant J'(\tau) \leqslant \Sigma_k' + O[K_1(\tau)]. \tag{24·2}$$

On inserting the value of Σ_0 from (23·9) it is seen that, when $\psi_0 = \frac{1}{2}$, $J'(\tau) \in C(\ln \tau^{-1}, 1)$, and, in fact,

$$J'(\tau) \to (2\psi_2)^{-\frac{1}{2}} \quad \text{as} \quad \tau \to \infty. \tag{24·3}$$

When $\psi_0 < \frac{1}{2}$, it follows from (24·2) and (23·11) that

$$J'(\tau) \in C(\ln \tau^{-1}, \exp k\tau).$$

24·1. By (23·22), the existence of all derivatives of $J(\tau)$ establishes the existence, when $\psi_0 = \frac{1}{2}$, of all derivatives of $f(\tau)$. In particular $f'(\tau) \in C(\ln \tau^{-1}, 1)$. Corresponding to Lemmas 19·2 and 19·3, we can now prove

Lemma 24·1. *When $\psi_0 = \frac{1}{2}$, $(2\psi_2)^{-\frac{1}{2}} f'(\tau)$ is the N-solution of the equation* (19·8).

Since $f(\tau)$ and $f'(\tau)$ satisfy the conditions of Theorem 15·3, it is easily verified that $\mathfrak{J}(\tau) = f'(\tau)/f(0)$ is a solution of (19·8). Since it is non-negative, it follows from Lemma 19·1 (i) that, if $\mathfrak{J}_N(\tau)$ is the N-solution,

$$f'(\tau)/f(0) - \mathfrak{J}_N(\tau)$$

is a non-negative solution of the homogeneous equation. It is not greater than $f'(\tau)/f(0)$ and therefore, by Theorem 19·2, it is $C(\ln \tau^{-1}, 1)$. Hence, by Theorem 22·1, it is zero. This proves the lemma, since $f(0) = (2\psi_2)^{\frac{1}{2}}$.

25. The Milne equation $(1 - \omega_0 \Lambda)_\tau \{J(t)\} = 0$

This equation is of such fundamental importance that it seems right to assemble the relevant formulae at this point. When $\Psi(x) = \frac{1}{2}\omega_0$, where $0 < \omega_0 \leqslant 1$,

$$T(\mu) = 1 - \tfrac{1}{2}\omega_0 \mu \log \frac{\mu + 1}{\mu - 1} \tag{25·1}$$

in the plane cut along $(-1, 1)$, $T(\mu)$ being the branch which is real when $\mu > 1$. Also

$$T_c(x) = 1 - \tfrac{1}{2}\omega_0 x \ln \frac{1 + x}{1 - x} \quad (-1 < x < 1). \tag{25·2}$$

When $0 < \omega_0 < 1$, the zeros of $T(\mu)$ will still be denoted by $\pm k^{-1}$ $(0 < k < 1)$.

When $\omega_0 = 1$, (12·12) and (12·15) give $h_0 = 1$, $h_1 = 1/\sqrt{3}$. After calculating h_2 by numerical integration, it is found from

(23·10) that†

$$\tau_0 = 0{\cdot}710,446,09 \tag{25·3}$$

correct to 8 decimal places. Hence Theorem 23·1 becomes‡

THEOREM 25·1. *Let $J(\tau)$ be the solution of*

$$(1-\Lambda)_\tau\{J(t)\} = 0 \tag{25·4}$$

in the class $C(1,\tau)$ such that $J(0) = 1$. Then $J(\tau)$ is given by (23·16), *where $k = 0$, $\Sigma_0 = \sqrt{3}(\tau+\tau_0)$, τ_0 is given by* (25·3), *$H(\mu)$ is the unique solution of*

$$\frac{1}{H(\mu)} = 1-\tfrac{1}{2}\mu\int_0^1 \frac{H(x)}{\mu+x}\,dx, \tag{25·5}$$

and $T_c(x)$ is given by (25·2) *($\omega_0 = 1$). If $j(s) = \mathfrak{L}_s\{J(\tau)\}$, then*

$$j(\mu^{-1}) = H(\mu) \quad (\mu \geqslant 0). \tag{25·6}$$

Any non-null solution of (25·4) *in the class $C(\ln\tau^{-1},\tau^{\alpha})$ $(\alpha \geqslant 1)$ is a constant multiple of this solution.*

When $\Psi(x) = \frac{1}{2}$, then $\psi_2 = \frac{1}{6}$ and equations (23·22) and (23·24) give

$$J(\tau) = \sqrt{3}f(\tau) = \sqrt{3}[\tau+q(\tau)], \tag{25·7}$$

and

$$f(0) = q(0) = 1/\sqrt{3}. \tag{25·8}$$

In the conservative case ($\omega_0 = 1$) one needs to know the solution for which the constant net flux is F. If $AJ(\tau)$ is the required solution, from (7.10)

$$F = \lim_{\tau\to\infty} 2A\tau^{-1}\int_0^\infty J(t)\,E_3(|\,t-\tau\,|)\,dt, \tag{25·9}$$

and hence from (25·7), by a simple calculation,§

$$F = \lim_{\tau\to\infty} 2\sqrt{3}A\tau^{-1}\left\{\tfrac{2}{3}\tau+E_5(\tau)+\int_0^\infty q(t)\,E_3(|\,t-\tau\,|)\,dt\right\}. \tag{25·10}$$

The integral on the right is bounded, since $q(t)$ is bounded for $t \geqslant 0$, and hence

$$F = 4A/\sqrt{3}. \tag{25·11}$$

Thus the required solution is

$$\tfrac{3}{4}F[\tau+q(\tau)]. \tag{25·12}$$

† Placzek and Seidel [1].
‡ There is no significant change in Theorem 23·2.
§ See Kourganoff [1], equation (14·9).

CHAPTER 6

SOLUTIONS OF THE NON-HOMOGENEOUS EQUATION

26. Introduction

In 1942 a mathematical method for solving the *homogeneous* Milne equation was published in Russian by V. A. Ambartsumian (see [1]). A second paper [2] dealt with the same problem from a physical point of view. The ideas of the second paper were developed by Chandrasekhar [3], who formulated 'principles of invariance' on which the solutions of transfer problems could be based. The application of these principles is not easy, and until a precise statement is given of the physical conditions which are sufficient to ensure their truth, any solution based on them ought to be verified in another way.†

An account of Ambartsumian's mathematical method was given in Kourganoff [1]. It depends upon the solution of an 'auxiliary equation' which leads directly to the H-equation. The technique by which this reduction is effected, together with the equations appearing in the reduction, provide a means by which the solutions of the *non-homogeneous* Milne equation can be found in terms of H-functions for certain special forms of the function $B(\tau)$. These forms happen to be the most important from a practical point of view.

The 'Ambartsumian technique' becomes more powerful when it is combined with the theory of N-solutions. In this chapter we develop this 'combined operations' method, and in Chaps. 7–9 it is used to develop the theory of finite atmospheres. Complex problems can often be solved by this method when the Wiener–Hopf technique fails.

The auxiliary equation for any given problem is the Milne equation obtained when the atmosphere is non-emitting and

† See, for example, Busbridge and Stibbs [1], Busbridge [1] and [4]. In Chandrasekhar's own work, the 'method of discrete ordinates' usually provides an independent check.

its surface is illuminated by radiation in a direction making a fixed angle $\cos^{-1}\mu_0$ $(0<\mu_0\leqslant 1)$ with the normal to the surface. Thus in § 5·1,

$$B_1(\tau)=0, \quad I_0(\mu')=\tfrac{1}{2}F\delta(\mu'-\mu_0), \tag{26·1}$$

where δ is Dirac's delta-function and F is a constant known as the 'intrinsic flux'. Then (5·12) gives

$$B(\tau)=\tfrac{1}{4}\omega_0 F\exp(-\tau/\mu_0), \tag{26·2}$$

and Milne's equation (5·11) becomes

$$\mathfrak{J}(\tau)=\omega_0\Lambda_\tau\{\mathfrak{J}(t)\}+\tfrac{1}{4}\omega_0 F\exp(-\tau/\mu_0). \tag{26·3}$$

On taking $F=4/\omega_0$ and writing $\mu_0=1/\sigma$ $(\sigma\geqslant 1)$, (26·3) becomes

$$(1-\omega_0\Lambda)_\tau\{\mathfrak{J}(t)\}=\exp(-\sigma\tau), \tag{26·4}$$

and this is the auxiliary equation.

When working with the generalized operator L, the corresponding auxiliary equation is

$$(1-L)_\tau\{\mathfrak{J}(t)\}=\exp(-\sigma\tau). \tag{26·5}$$

27. The auxiliary equation (26·5)

By the corollary to Theorem 19·4 when $\psi_0=\frac{1}{2}$ and $\sigma>0$, and by Theorem 19·5 when $\psi_0<\frac{1}{2}$ and $\sigma\geqslant 0$, the N-series of the equation (26·5) converges to the N-solution. We shall denote the N-solution by $J(\tau,\sigma)$, so that†

$$J(\tau,\sigma)=\sum_{\nu=0}^{\infty}L_\tau^\nu\{\exp(-\sigma t)\}, \tag{27·1}$$

and

$$(1-L)_\tau\{J(t,\sigma)\}=\exp(-\sigma\tau), \tag{27·2}$$

with the restrictions $\sigma>0$ when $\psi_0=\frac{1}{2}$, $\sigma\geqslant 0$ when $\psi_0<\frac{1}{2}$.

27·1. *Properties of $J(\tau,\sigma)$.* The main difficulty in the rigorous treatment of Ambartsumian's reduction of the auxiliary equation to that for $H(\mu)$ is the proof of the existence of $(\partial/\partial\tau)J(\tau,\sigma)$

† In the probabilistic theory of Sobolev [1]–[5] and Ueno [1]–[4],

$$J(\tau,\mu^{-1})=p(\mu,\tau),$$

where $p(\mu,\tau)\,d\mu$ is the probability that a photon absorbed at the depth τ will be re-emitted in a direction lying between μ and $\mu+d\mu$.

when $\psi_0 = \frac{1}{2}$† and the investigation of its behaviour for large and small τ. For this reason we begin with the following theorem:‡

THEOREM 27·1. *When $\psi_0 = \frac{1}{2}$ let $\sigma > 0$, and when $\psi_0 < \frac{1}{2}$ let $\sigma \geqslant 0$. Then the N-solution $J(\tau, \sigma)$ of (26·5) can be written in the form*

$$J(\tau,\sigma) = J(0,\sigma)\left\{J(\tau)-(k+\sigma)\int_0^\tau J(t)\exp\left[-\sigma(\tau-t)\right]dt\right\}, \quad (27\cdot 3)$$

where $J(\tau)$ is the solution of $(1-L)_\tau\{J(t)\} = 0$ which is given by (23·16) and k^{-1} is the non-negative root of $T(\mu) = 0$ ($k = 0$ when $\psi_0 = \frac{1}{2}$).

First suppose that $\sigma > 0$ and let

$$\mathfrak{J}(\tau) = \int_0^\tau J(t)\exp\left[-\sigma(\tau-t)\right]dt. \quad (27\cdot 4)$$

We shall show that this is a solution of

$$(1-L)_\tau\{\mathfrak{J}(t)\} = a\exp(-\sigma\tau) \quad (27\cdot 5)$$

for some a. From (27·4) we have

$$\mathfrak{J}'(\tau)+\sigma\mathfrak{J}(\tau) = J(\tau), \quad \mathfrak{J}(0) = 0. \quad (27\cdot 6)$$

If $\psi_0 < \frac{1}{2}$, it follows from (23·16) and (23·11) that

$$J(\tau) = A\exp(k\tau)+O(1), \quad (27\cdot 7)$$

where $A = \frac{1}{2}c_k\exp(k\tau_k)$ and $O(1)$ represents a function of $C(1,1)$. From (27·4) and (27·7), since $\sigma > 0$,

$$\mathfrak{J}(\tau) = A(k+\sigma)^{-1}\exp(k\tau)+O(1) \quad (27\cdot 8)$$

and therefore, by (27·6),

$$\mathfrak{J}'(\tau) = Ak(k+\sigma)^{-1}\exp(k\tau)+O(1). \quad (27\cdot 9)$$

Hence $\mathfrak{J}(\tau)$ satisfies the conditions of Theorem 15·3, and if

$$\phi(\tau) = \mathfrak{J}(\tau)-L_\tau\{\mathfrak{J}(t)\}, \quad (27\cdot 10)$$

then (since $\mathfrak{J}(0) = 0$)

$$\phi'(\tau) = \mathfrak{J}'(\tau)-L_\tau\{\mathfrak{J}'(t)\}. \quad (27\cdot 11)$$

† When $\psi_0 < \frac{1}{2}$ no difficulty arises. See Busbridge [2], where a direct proof is given.

‡ See Hopf [1] when $L = \Lambda$. This is a case in which $\psi_0 = \frac{1}{2}$.

On adding σ times (27·10) to (27·11) and using (27·6), we get

$$\phi'(\tau)+\sigma\phi(\tau) = J(\tau)-L_\tau\{J(t)\} = 0. \tag{27·12}$$

Hence $\phi(\tau) = a\exp(-\sigma\tau)$ for some a, and $\mathfrak{J}(\tau)$ is a solution of (27·5). If $a = 0$, then $\mathfrak{J}(\tau) = BJ(\tau)$ for some constant B; and $B = 0$ since $\mathfrak{J}(0) = 0, J(0) = 1$. From (27·6), $\mathfrak{J}(\tau) = 0$ implies $J(\tau) = 0$, which is not so, and hence $a \neq 0$.

The function $\mathfrak{J}(\tau)+BJ(\tau)$, where B is a constant, is also a solution of (27·5). By (27·7) and (27·8)

$$\mathfrak{J}(\tau)+BJ(\tau) = A[(k+\sigma)^{-1}+B]\exp(k\tau)+O(1) = O(1) \tag{27·13}$$

if $B = -(k+\sigma)^{-1}$. Hence

$$\mathfrak{J}(\tau)-(k+\sigma)^{-1}J(\tau)\in C(1,1),$$

and therefore, by Lemma 19·1 (ii), it is the N-solution of (27·5), i.e. it is $aJ(\tau,\sigma)$, and hence

$$J(\tau,\sigma) = a^{-1}[\mathfrak{J}(\tau)-(k+\sigma)^{-1}J(\tau)]. \tag{27·14}$$

Since $\mathfrak{J}(0) = 0$ and $J(0) = 1$, therefore

$$a^{-1} = -(k+\sigma)J(0,\sigma) \tag{27·15}$$

and (27·3) follows.

When $\psi_0<\frac{1}{2}$ and $\sigma = 0$, the same analysis holds except that, in deducing (27·8) from (27·4), it is necessary to use the more precise equation (23·18) in place of (27·7).

When $\psi_0 = \frac{1}{2}$ and $\sigma>0$, the proof is similar except that, in (27·7)–(27·9) and in (27·13), $\exp(k\tau)$ is replaced by τ. Elsewhere $k = 0$.

From this theorem and § 24 we can deduce

THEOREM 27·2. *When $\psi_0 = \frac{1}{2}$, let $\sigma>0$, and when $\psi_0<\frac{1}{2}$, let $\sigma\geqslant 0$. Then $J(\tau,\sigma)$ is the unique solution of* (26·5) *in the class $C(1,1)$. The derivative $(\partial/\partial\tau)J(\tau,\sigma)$ exists for $\tau>0$ and is $C(\ln\tau^{-1},1)$.*

The properties of $J(\tau,\sigma)$ follow from the preceding proof. When $\psi_0<\frac{1}{2}$ and $\sigma>0$, $\mathfrak{J}'(\tau)\in C(1,\exp k\tau)$ by (27·9). From (27·14) and (27·15), $(\partial/\partial\tau)J(\tau,\sigma)$ exists and

$$(\partial/\partial\tau)J(\tau,\sigma) = J(0,\sigma)[J'(\tau)-(k+\sigma)\mathfrak{J}'(\tau)].$$

By § 24, $J'(\tau) \in C(\ln \tau^{-1}, \exp k\tau)$ and hence $(\partial/\partial\tau) J(\tau,\sigma)$ is $O(\ln \tau^{-1})$ as $\tau \to +0$. For large τ, by (24·2) and (27·9),

$$(\partial/\partial\tau) J(\tau,\sigma) = AJ(0,\sigma)(k-k)\exp(k\tau) + O(1) = O(1).$$

When $\psi_0 < \frac{1}{2}$ and $\sigma = 0$ or when $\psi_0 = \frac{1}{2}$ and $\sigma > 0$, the proof has to be modified as indicated above and the same results are found to hold.

27·2. Lemma 27·1. *If $\tau \geqslant 0$, $J(\tau,\sigma)$ is a continuous function of σ for $\sigma > 0$; and if $\tau > 0$, $J(\tau,\sigma) \to 0$ as $\sigma \to \infty$, but $J(0,\sigma) \to 1$ as $\sigma \to \infty$.*

Let $0 < \sigma < \sigma_1$; then $\exp(-\sigma\tau) \geqslant \exp(-\sigma_1\tau)$ and hence, by (27·1),

$$J(t,\sigma) \geqslant J(t,\sigma_1).$$

As σ increases to σ_1, $J(t,\sigma)$ steadily decreases to a bounded function $\bar{J}(t,\sigma_1)$. By a simple extension of the limit theorem for monotonic sequences,

$$\lim_{\sigma \to \sigma_1 - 0} L_\tau\{J(t,\sigma)\} = L_\tau\left\{\lim_{\sigma \to \sigma_1 - 0} J(t,\sigma)\right\} = L_\tau\{\bar{J}(t,\sigma_1)\}$$

for any fixed $\tau \geqslant 0$.

When σ_1 is finite and $\tau \geqslant 0$, on letting $\sigma \to \sigma_1 - 0$ in (27·2), we get

$$\bar{J}(\tau,\sigma_1) = L_\tau\{\bar{J}(t,\sigma_1)\} + \exp(-\tau\sigma_1). \tag{27·16}$$

Since $\bar{J}(\tau,\sigma_1)$ is bounded, it is $C(1,1)$ (by Theorem 19·2) and hence

$$\bar{J}(\tau,\sigma_1) = J(\tau,\sigma_1)$$

(by Theorem 27·2). Thus $J(\tau,\sigma) \to J(\tau,\sigma_1)$ as $\sigma \to \sigma_1 - 0$. Similarly, $J(\tau,\sigma)$ is continuous at σ_1 on the right.

When $\tau > 0$ and σ_1 is infinite, (27·16) is the homogeneous equation and $\bar{J}(\tau,\infty) = 0$ by Theorems 22·1 and 22·2.

When $\tau = 0$ and $\sigma \to \infty$, we get in place of (27·16)

$$\bar{J}(0,\infty) = L_0\{\bar{J}(t,\infty)\} + 1. \tag{27·17}$$

But $\bar{J}(t,\infty) = 0$ for $t > 0$ and so $\bar{J}(0,\infty) = 1$.

Lemma 27·2.

$$\int_1^\infty J(\tau,x)\Psi(x^{-1})x^{-1}\,dx$$

is the N-solution of

$$(1-L)_\tau\{\mathfrak{J}(t)\} = K_1(\tau) \tag{27·18}$$

and it is $C(\ln \tau^{-1}, 1)$.

Since the terms in (27·1) are positive and the integrands of all the repeated integrals are positive,

$$\int_1^\infty J(\tau, x)\,\Psi(x^{-1})\,x^{-1}\,dx = \sum_{\nu=0}^{\infty} L_\tau^\nu\left\{\int_1^\infty \exp(-xt)\,\Psi(x^{-1})\,x^{-1}\,dx\right\}$$

$$= \sum_{\nu=0}^{\infty} L_\tau^\nu\{K_1(t)\} \qquad (27{\cdot}19)$$

if this series is convergent. When $\psi_0 = \frac{1}{2}$ it converges to $(2\psi_2)^{-\frac{1}{2}} f'(\tau)$ by Lemma 24·1; it is the N-solution and it is $C(\ln \tau^{-1}; 1)$ (see § 24).

When $\psi_0 < \frac{1}{2}$, since $K_1(\tau) \in C(\ln \tau^{-1}, 1)$, it follows from Theorem 15·1 that there is a constant M such that

$$L_\tau\{K_1(t)\} \leqslant M \quad (\tau \geqslant 0),$$

and hence, by (15·6), that

$$L_\tau^\nu\{K_1(t)\} \leqslant M(2\psi_0)^{\nu-1} \quad (\nu \geqslant 1).$$

Thus the series (27·19) converges and its sum is less than

$$K_1(\tau) + M/(1-2\psi_0).$$

From this the order conditions follow. Since $K_1(\tau) \geqslant 0$, the sum of the N-series is the N-solution (Theorem 19·1); and it is continuous for $\tau > 0$ (Theorem 19·2).

28. The reduction to the H-equation

Ambartsumian's technique will be used in proving the following theorem. The successive steps, which form the technique, will be found to be repeated in the solutions of §§ 29·1 and 29·2.

THEOREM 28·1. *If $J(\tau, \sigma)$ is the N-solution of* (26·5) *and if, for $\sigma > 0$, $s > 0$,*

$$R(s, \sigma) = \int_0^\infty J(\tau, \sigma)\exp(-s\tau)\,d\tau, \qquad (28{\cdot}1)$$

then

$$R(s, \sigma) = H(s^{-1})\,H(\sigma^{-1})/(s+\sigma), \qquad (28{\cdot}2)$$

and

$$J(0, \mu^{-1}) = H(\mu) \quad (\mu \geqslant 0). \qquad (28{\cdot}3)$$

Since, by Theorem 27·2, $J(\tau, \sigma)$ satisfies the conditions of Theorem 15.3, we may differentiate (27·2), giving

$$(1-L)_\tau\left\{\frac{\partial}{\partial t} J(t, \sigma)\right\} = J(0, \sigma)\,K_1(\tau) - \sigma\exp(-\sigma\tau). \qquad (28{\cdot}4)$$

Adding σ times (27·2), we have

$$(1-L)_\tau\left\{\frac{\partial}{\partial t}J(t,\sigma)+\sigma J(t,\sigma)\right\} = J(0,\sigma)\,K_1(\tau). \qquad (28\cdot5)$$

By Lemma 27·2,

$$(1-L)_\tau\left\{\int_1^\infty J(t,x)\,\Psi(x^{-1})\frac{dx}{x}\right\} = K_1(\tau), \qquad (28\cdot6)$$

and hence

$$(1-L)_\tau\left\{\frac{\partial}{\partial t}J(t,\sigma)+\sigma J(t,\sigma)-J(0,\sigma)\int_1^\infty J(t,x)\,\Psi(x^{-1})\frac{dx}{x}\right\} = 0. \qquad (28\cdot7)$$

By Theorem 27·2 and Lemma 27·2, the function in curled brackets is $C(\ln t^{-1}, 1)$ and hence it is zero by Theorems 22·1 and 22·2. Thus†

$$\frac{\partial}{\partial \tau}J(\tau,\sigma)+\sigma J(\tau,\sigma) = J(0,\sigma)\int_1^\infty J(\tau,x)\,\Psi(x^{-1})\frac{dx}{x}. \qquad (28\cdot8)$$

Let $R(s,\sigma)$ be defined by (28·1). By Theorem 19·3 applied to (27·2) and the similar equation with s in place of σ,

$$\int_0^\infty J(\tau,\sigma)\exp(-s\tau)\,d\tau = \int_0^\infty J(\tau,s)\exp(-\sigma\tau)\,d\tau,$$

the integrals converging for $\sigma>0, s>0$. Hence

$$R(s,\sigma) = R(\sigma,s). \qquad (28\cdot9)$$

Multiply (28·8) by $\exp(-s\tau)$ $(s>0)$ and integrate with respect to τ over $(0,\infty)$, integrating the first term by parts. Then‡

$$-J(0,\sigma)+sR(s,\sigma)+\sigma R(s,\sigma) = J(0,\sigma)\int_1^\infty R(s,x)\,\Psi(x^{-1})\frac{dx}{x},$$

i.e.
$$(s+\sigma)\,R(s,\sigma) = J(0,\sigma)\left\{1+\int_1^\infty R(s,x)\,\Psi(x^{-1})\frac{dx}{x}\right\}. \qquad (28\cdot10)$$

† Equation (28·8) corresponds to the stochastic integro-differential equation of the probabilistic theory. (See the introduction and the footnote on p. 63.)

‡ The inversion of the order of integration here, and subsequently, is justified by absolute convergence.

Put $\tau = 0$ and write s for σ in (27·2). Then

$$J(0,s) = \int_0^\infty J(t,s)\,K_1(t)\,dt + 1$$
$$= 1 + \int_0^\infty J(t,s)\,dt \int_1^\infty \Psi(x^{-1})\exp(-xt)\frac{dx}{x}$$
$$= 1 + \int_1^\infty \Psi(x^{-1})\,R(x,s)\frac{dx}{x}$$
$$= 1 + \int_1^\infty \Psi(x^{-1})\,R(s,x)\frac{dx}{x} \qquad (28\cdot 11)$$

by (28·9). Hence (28·10) becomes

$$(s+\sigma)\,R(s,\sigma) = J(0,\sigma)\,J(0,s). \qquad (28\cdot 12)$$

From (28·11) and (28·12),

$$J(0,s) = 1 + J(0,s)\int_1^\infty \frac{\Psi(x^{-1})\,J(0,x)}{x(s+x)}\,dx.$$

On putting $s = 1/\mu$, $x = 1/u$, we get

$$J(0,\mu^{-1}) = 1 + \mu J(0,\mu^{-1})\int_0^1 \frac{\Psi(u)\,J(0,u^{-1})}{\mu+u}\,du. \qquad (28\cdot 13)$$

Hence $J(0,\mu^{-1})$ is a solution of (8·1). By Lemma 27·1 it is continuous for $\mu \geqslant 0$ and hence, by Theorem 11·1,

$$J(0,\mu^{-1}) = H(\mu).$$

This proves (28·3) and (28·2) follows from (28·12).

From (28·3) and (27·1), we have

COROLLARY.

$$H(\mu) = \left[\sum_{\nu=0}^{\infty} L_\tau^\nu\{\exp(-t/\mu)\}\right]_{\tau=0} \qquad (\mu > 0). \qquad (28\cdot 14)$$

28·1. *The scattering function.* Chandrasekhar, in [1], makes extensive use of a 'scattering function'. This is the function $S(\mu,\mu')$ defined by

$$S(\mu,\mu') = R(1/\mu, 1/\mu'). \qquad (28\cdot 15)$$

(Cf. loc. cit. p. 85, equation (120). Chandrasekhar inserts ω_0 as a multiplier, but this is better omitted as it makes generalizations difficult.)

A consideration of the use of the scattering function for finding the emergent intensity will be deferred until § 49, but the relevant sections (45–47 and 49) could be inserted here.

29. The solution of

$$(1-L)_\tau\{\mathfrak{J}(t)\} = \sum_{\nu=0}^{n} a_\nu \tau^{n-\nu}/(n-\nu)! \tag{29·1}$$

When $\psi_0 < \frac{1}{2}$, the N-solution of (29·1) exists and is $C(1, \tau^n)$ (by Theorem 19·6). When $\psi_0 = \frac{1}{2}$, consider the case in which every $a_\nu \geqslant 0$. Then the N-solution (if it exists) is non-negative. But the polynomial is not $L(0, \infty)$ and therefore, by Theorem 19·4, there is no non-negative solution; hence the solution found below in § 29·2 is not the N-solution.

The Ambartsumian technique is used in §§ 29·1 and 29·2. In § 29·3 a different method of obtaining the 'emergent intensity' is considered (see Ueno [2]).

We shall use $J_n(\tau)$ to denote a solution of (29·1), so that

$$(1-L)_\tau\{J_n(t)\} = \sum_{\nu=0}^{n} a_\nu \tau^{n-\nu}/(n-\nu)!, \tag{29·2}$$

and we shall consider the sequence of equations given by $n = 0, 1, \ldots$.

29·1. *The case* $\psi_0 < \frac{1}{2}$. Let $J_n(\tau)$ be the N-solution, so that $J_n(\tau) \in C(1, \tau^n)$. We shall assume that $J'_n(\tau)$ exists and is $C(\ln \tau^{-1}, \tau^n)$. This will be verified later.

By Theorem 15·3, (29·2) can be differentiated, giving

$$(1-L)_\tau\{J'_n(t)\} = J_n(0)\,K_1(\tau) + \sum_{\nu=0}^{n-1} a_\nu \tau^{n-\nu-1}/(n-\nu-1)!. \tag{29·3}$$

When $n = 0$, the polynomial terms are missing. From (29·3) and (29·2),

$$(1-L)_\tau\{J'_n(t) - J_{n-1}(t)\} = J_n(0)\,K_1(\tau), \tag{29·4}$$

and this holds for $n = 0$ if $J_{-1}(\tau)$ is defined to be zero. On subtracting $J_n(0)$ times (28·6) from (29·4), we get

$$(1-L)_\tau\left\{J'_n(t) - J_{n-1}(t) - J_n(0)\int_1^\infty J(t,x)\,\Psi(x^{-1})\frac{dx}{x}\right\} = 0. \tag{29·5}$$

The function in curled brackets is, by hypothesis and Lemma 27·2, $C(\ln\tau^{-1}, \tau^n)$, and hence, by Theorem 22·2, it is zero. Thus

$$J'_n(\tau) - J_{n-1}(\tau) = J_n(0)\int_1^\infty J(\tau, x)\,\Psi(x^{-1})\frac{dx}{x}. \qquad (29\cdot6)$$

Let
$$j_n(s) = \mathfrak{L}_s\{J_n(\tau)\} \quad (s > 0), \qquad (29\cdot7)$$

where $\mathfrak{L}_s$ is defined by (5·16). Operate on (29·6) by $s^{-1}\mathfrak{L}_s$, integrating the first term by parts. Then, by (28·1),

$$-J_n(0) + j_n(s) - s^{-1}j_{n-1}(s) = J_n(0)\int_1^\infty R(s, x)\,\Psi(x^{-1})\frac{dx}{x},$$

and hence, by (28·11) and (28·3),

$$j_n(s) - s^{-1}j_{n-1}(s) = J_n(0)\,J(0, s) = J_n(0)\,H(s^{-1}). \qquad (29\cdot8)$$

Since $j_{-1}(s) = 0$, on solving for $j_n(s)$ and writing $s = 1/\mu$, we get

$$j_n(\mu^{-1}) = H(\mu)\,[J_n(0) + \mu J_{n-1}(0) + \ldots + \mu^n J_0(0)]. \qquad (29\cdot9)$$

By means of (29·9), the conditions assumed for $J_n(\tau)$ are easily verified. Since $\psi_0 < \frac{1}{2}$, we can let $\sigma \to 0$ in (28·1) and (28·2). From (12·9) and (12·4),

$$1/H(\mu) \to 1 - h_0 = (1 - 2\psi_0)^{\frac{1}{2}} \quad \text{as} \quad \mu \to \infty,$$

and hence (28·1) and (28·2) give

$$\mathfrak{L}_s\{J(\tau, 0)\} = sR(s, 0) = (1 - 2\psi_0)^{-\frac{1}{2}}\,H(s^{-1}). \qquad (29\cdot10)$$

The functions whose Laplace transforms are $H(s^{-1})$, $s^{-1}H(s^{-1})$, $s^{-2}H(s^{-1})$, etc., are therefore constant multiples of

$$J(\tau, 0), \quad \int_0^\tau J(t, 0)\,dt, \quad \int_0^\tau dt\int_0^t J(t', 0)\,dt', \quad \text{etc.} \qquad (29\cdot11)$$

By Theorem 27·2, all these functions satisfy the assumed conditions and therefore $J_n(\tau) \in C(1, \tau^n)$ and $J'_n(\tau) \in C(\ln\tau^{-1}, \tau^n)$.

The constants $J_0(0), J_1(0), \ldots$ remain to be found. From (29·2), with $\tau = 0$,

$$J_n(0) = \int_0^\infty J_n(t)\,K_1(t)\,dt + a_n \quad (n = 0, 1, \ldots). \qquad (29\cdot12)$$

On replacing $K_1(t)$ by its definition, we get

$$\begin{aligned}
J_n(0) &= \int_0^\infty J_n(t)\,dt \int_1^\infty \Psi(x^{-1}) \exp(-xt)\frac{dx}{x} + a_n \\
&= \int_1^\infty \Psi(x^{-1})\frac{dx}{x}\int_0^\infty J_n(t)\exp(-xt)\,dt + a_n \\
&= \int_1^\infty \Psi(x^{-1})\, j_n(x)\frac{dx}{x^2} + a_n \\
&= \int_0^1 \Psi(\mu)\, j_n(\mu^{-1})\,d\mu + a_n. \qquad (29\cdot13)
\end{aligned}$$

Hence (29·9) and (29·13) give

$$J_n(0) = \sum_{\nu=0}^{n} h_\nu J_{n-\nu}(0) + a_n \quad (n = 0, 1, \ldots), \qquad (29\cdot14)$$

where h_ν is defined by (12·2). By (12·4), this can be written

$$(1-2\psi_0)^{\frac{1}{2}} J_n(0) - \sum_{\nu=1}^{n} h_\nu J_{n-\nu}(0) = a_n \quad (n = 0, 1, \ldots). \quad (29\cdot15)$$

Solving, we have, for $\nu = 0, 1, 2, \ldots,$

$$J_\nu(0) = (1-2\psi_0)^{-\frac{1}{2}(\nu+1)} \begin{vmatrix} a_\nu & -h_1 & \ldots & -h_\nu \\ a_{\nu-1} & (1-2\psi_0)^{\frac{1}{2}} & \ldots & -h_{\nu-1} \\ a_{\nu-2} & 0 & \ldots & -h_{\nu-2} \\ \ldots & \ldots & \ldots & \ldots \\ a_0 & 0 & \ldots & (1-2\psi_0)^{\frac{1}{2}} \end{vmatrix}. \qquad (29\cdot16)$$

Because the existence of $J'_n(\tau)$ was initially unknown, it is necessary to verify that the functions $J_n(\tau)$, whose Laplace transforms are given by (29·9) and (29·16), are solutions of (29·1). It is not difficult to see that the argument is reversible.† Thus (29·16) implies (29·15) and hence (by (12·4)) (29·14). This and (29·9) together give (29·13), from which (29·12) follows. Thus we have to show that (29·12) and (29·9) together imply (29·2) ($n = 0, 1, \ldots$).

† The method of § 22 can be used, but the analysis is heavy.

The steps leading to (29·9) can be reversed up to (29·4), $j_{-1}(s)$ and hence $J_{-1}(\tau)$ being defined to be zero. If

$$\phi_n(\tau) = (1-L)_\tau \{J_n(t)\} \quad (n = -1, 0, 1, \ldots), \tag{29·17}$$

then (29·4) can be written

$$\phi_n'(\tau) = \phi_{n-1}(\tau) \quad (n = 0, 1, \ldots), \tag{29·18}$$

and hence $$\phi_n^{(n+1)}(\tau) = \phi_{-1}(\tau) = 0.$$

Thus $$\phi_n(\tau) = \sum_{\nu=0}^{n} b_\nu \tau^{n-\nu}/(n-\nu)!$$

for some constants $b_0, \ldots, b_n$, and (29·17) gives

$$(1-L)_\tau \{J_n(t)\} = \sum_{\nu=0}^{n} b_\nu \tau^{n-\nu}/(n-\nu)!. \tag{29·19}$$

On putting $\tau = 0$ in (29·19) and comparing with (29·12), we get $b_n = a_n$ $(n = 0, 1, \ldots)$. Hence $J_n(\tau)$ satisfies (29·2). Thus we have proved†

THEOREM 29·1. *When $\psi_0 < \frac{1}{2}$, the equation* (29·1) *has a solution $J_n(\tau)$ which is $C(1, \tau^n)$. This is the N-solution. If $j_n(s) = \mathfrak{L}_s\{J_n(\tau)\}$, then $j_n(\mu^{-1})$ $(\mu \geq 0)$ is given by* (29·9) *and* (29·16).

29·2. *The case $\psi_0 = \frac{1}{2}$.* The above solution breaks down when $\psi_0 = \frac{1}{2}$ because the solution of the homogeneous equation is not zero if it is $O(\tau^n)$ for large τ and therefore (29·6) cannot be deduced from (29·5).

We shall look for solutions $J_n(\tau)$ which are $C(1, \tau^{\alpha_n})$ for some $\alpha_n \geq 1$ and whose first derivatives are $C(\ln \tau^{-1}, \tau^{\alpha_n})$. Then (29·6) has to be replaced by

$$J_n'(\tau) - J_{n-1}(\tau) = J_n(0) \int_1^\infty J(\tau, x)\, \Psi(x) \frac{dx}{x} + A_n J(\tau), \tag{29·20}$$

where A_n is a constant and $J(\tau)$ is given by (23·16) with $k = 0$. Then, by (23·19),

$$j(s) = \mathfrak{L}_s\{J(\tau)\} = H(s^{-1}). \tag{29·21}$$

Operating on (29·20) by $s^{-1}\mathfrak{L}_s$, we get

$$-J_n(0) + j_n(s) - s^{-1} j_{n-1}(s)$$
$$= J_n(0) \int_1^\infty R(s, x)\, \Psi(x^{-1}) \frac{dx}{x} + A_n s^{-1} j(s),$$

† The results of this section were first found formally by S. S. Huang [1] by guessing the form for $j_n(s)$.

and hence by (28·11) and (28·3)

$$\begin{aligned} j_n(s) - s^{-1} j_{n-1}(s) &= J_n(0)\, J(0, s) + A_n\, s^{-1} j(s) \\ &= J_n(0)\, H(s^{-1}) + A_n\, s^{-1} H(s^{-1}). \qquad (29\cdot 22) \end{aligned}$$

This gives (since $j_{-1}(s) = 0$)

$$\begin{aligned} j_n(\mu^{-1}) &= H(\mu)\, \{J_n(0) + \mu J_{n-1}(0) + \dots + \mu^n J_0(0) \\ &\qquad + \mu A_n + \mu^2 A_{n-1} + \dots + \mu^{n+1} A_0\} \\ &= H(\mu)\, \{J_n(0) + A'_n\, \mu + A'_{n-1}\, \mu^2 + \dots + A'_0\, \mu^{n+1}\}, \qquad (29\cdot 23) \end{aligned}$$

where $\quad A'_0 = A_0, \quad A'_\nu = A_\nu + J_{\nu-1}(0) \quad (\nu \geqslant 1). \qquad (29\cdot 24)$

Equation (29·13) is unchanged, and on substituting for $j_n(\mu^{-1})$ from (29·23), we get

$$J_n(0) = J_n(0)\, h_0 + \sum_{\nu=1}^{n+1} h_\nu\, A'_{n+1-\nu} + a_n.$$

But $h_0 = 1$ when $\psi_0 = \frac{1}{2}$ and hence

$$A'_n\, h_1 + A'_{n-1}\, h_2 + \dots + A'_0\, h_{n+1} = -a_n \quad (n = 0, 1, \dots). \qquad (29\cdot 25)$$

Solving, we have for $\nu = 0, 1, \dots,$

$$A'_\nu = -h_1^{-(\nu+1)} \begin{vmatrix} a_\nu & h_2 & \dots & h_{\nu+1} \\ a_{\nu-1} & h_1 & \dots & h_\nu \\ a_{\nu-2} & 0 & \dots & h_{\nu-1} \\ \dots & \dots & \dots & \dots \\ a_0 & 0 & \dots & h_1 \end{vmatrix}. \qquad (29\cdot 26)$$

Equations (29·23) and (29·26) give $j_n(\mu^{-1})$, $J_n(0)$ being arbitrary.

On comparing (29·10) with (29·21), it is seen that the verification of the conditions assumed for $J_n(\tau)$ will be similar to that given in § 29·1, but with $J(\tau)$ in place of $J(\tau, 0)$ and with Theorem 23·1 and § 24 in place of Theorem 27·2. Since $J(\tau) \in C(1, \tau)$, it is found that $J_n(\tau) \in C(1, \tau^{n+2})$.

As in § 29·1, by reversing the proof it can be verified that the functions $J_n(\tau)$, whose Laplace transforms are given by (29·23) and (29·26), do satisfy (29·1) $(n = 0, 1, \dots)$ for any values of $J_n(0)$. Thus we have

THEOREM 29·2. *When* $\psi_0 = \frac{1}{2}$, *the equation* (29·1) *has a solution* $J_n(\tau)$ *which is* $C(1, \tau^{n+2})$. *If* $j_n(s) = \mathfrak{L}_s\{J_n(\tau)\}$, *then* $j_n(\mu^{-1})$ $(\mu \geqslant 0)$ *is given by* (29·23) *and* (29·26), $J_n(0)$ *being arbitrary.*

It is of particular interest to note that the equation

$$(1-L)_\tau\{\mathfrak{J}(t)\} = 1 \tag{29·27}$$

has, when $\psi_0 = \frac{1}{2}$, the solution given by

$$j_0(s) = -h_1^{-1}s^{-1}H(s^{-1}), \quad J_0(\tau) = -h_1^{-1}\int_0^\tau J(t)\,dt, \tag{29·28}$$

and therefore, by (23·22), since $h_1 = \surd(2\psi_2)$,

$$\mathfrak{J}(\tau) = J_0(\tau) = -(2\psi_2)^{-1}\int_0^\tau f(t)\,dt, \tag{29·29}$$

where $f(\tau)$ is the function of Theorem 17·1. This can be verified directly by integrating the equation $(1-L)_\tau\{f(t)\} = 0$ over the interval $(0, \tau)$. The solution is everywhere negative.

29·3. *Ueno's method.* A simple formula for the emergent intensity corresponding to the N-solution of the equation

$$(1-L)_\tau\{\mathfrak{J}(t)\} = B(\tau) \tag{29·30}$$

has been found by Ueno [2]. It is given in

THEOREM 29·3. *Let* $B(\tau) \geqslant 0$ *and let* $\mathfrak{J}_N(\tau)$ *be the* N*-solution of* (29·30). *Then*

$$\mathfrak{j}_N(\mu^{-1}) = \mathfrak{L}_{1/\mu}\{\mathfrak{J}_N(\tau)\} = \int_0^\infty J(\tau, \mu^{-1})\,B(\tau)\frac{d\tau}{\mu}, \tag{29·31}$$

where $J(\tau, \sigma)$ *is the* N*-solution of* (26·5).

The proof follows at once from Theorem 19·3 which, applied to equations (29·30) and (26·5), gives

$$\int_0^\infty \mathfrak{J}_N(\tau)\exp(-\sigma\tau)\,d\tau = \int_0^\infty J(\tau, \sigma)\,B(\tau)\,d\tau.$$

When this theorem is applied to (29·1), we get

$$\begin{aligned} j_n(\mu^{-1}) &= \sum_{\nu=0}^{n} \frac{a_\nu \mu^{-1}}{(n-\nu)!}\int_0^\infty J(t, \mu^{-1})\,t^{n-\nu}\,dt \\ &= \sum_{\nu=0}^{n} a_\nu \mu^{-1}\phi_{n-\nu}(\mu)/(n-\nu)\,!, \end{aligned} \tag{29·32}$$

where $$\phi_m(\mu) = \int_0^\infty J(t,\mu^{-1})\, t^m\, dt \quad (m \geqslant 0). \tag{29·33}$$

By a more detailed investigation of the order conditions in Theorem 27·1, it can be shown that, when $\psi_0 < \frac{1}{2}$ and $\mu > 0$,

$$t^m J(t,\mu^{-1}) \to 0 \quad \text{as} \quad t \to \infty$$

for all $m \geqslant 0$. Assuming this, integrating (29·33) by parts, and using equations (28·8) and (28·3), we obtain the relation

$$\mu^{-1}\phi_m(\mu) - m\phi_{m-1}(\mu) = H(\mu)\int_0^1 t^{-1}\phi_m(t)\,\Psi(t)\,dt \quad (m \geqslant 1). \tag{29·34}$$

This can be transformed into a reduction formula for $\phi_m(\mu)$ as follows: Multiply (29·34) by $\Psi(\mu)$ and integrate over $(0, 1)$. Then

$$\int_0^1 t^{-1}\phi_m(t)\,\Psi(t)\,dt - m\int_0^1 \phi_{m-1}(t)\,\Psi(t)\,dt = h_0\int_0^1 t^{-1}\phi_m(t)\,\Psi(t)\,dt.$$

Since $1 - h_0 = (1-2\psi_0)^{\frac{1}{2}}$,

$$\int_0^1 t^{-1}\phi_m(t)\,\Psi(t)\,dt = m(1-2\psi_0)^{-\frac{1}{2}}\int_0^1 \phi_{m-1}(t)\,\Psi(t)\,dt, \tag{29·35}$$

and (29·34) becomes

$$\phi_m(\mu) = m\mu\left[\phi_{m-1}(\mu) + (1-2\psi_0)^{-\frac{1}{2}} H(\mu)\int_0^1 \phi_{m-1}(t)\,\Psi(t)\,dt\right]. \tag{29·36}$$

It remains to find $\phi_0(\mu)$. From the equation

$$\int_0^\infty (\partial/\partial\tau)\, J(\tau,\mu^{-1})\,d\tau = -J(0,\mu^{-1}) = -H(\mu) \tag{29·37}$$

and (28·8), we obtain

$$\mu^{-1}\phi_0(\mu) - H(\mu) = H(\mu)\int_0^1 t^{-1}\phi_0(t)\,\Psi(t)\,dt. \tag{29·38}$$

Again multiplying by $\Psi(\mu)$ and integrating over $(0, 1)$, we get

$$(1-h_0)\int_0^1 t^{-1}\phi_0(t)\,\Psi(t)\,dt = h_0, \tag{29·39}$$

and hence $$\phi_0(\mu) = (1-2\psi_0)^{-\frac{1}{2}}\,\mu H(\mu). \tag{29·40}$$

From (29·36) and (29·40) it is found that

$$\phi_m(\mu) = m!\,\mu H(\mu) Q_m(\mu), \tag{29·41}$$

where

$$Q_m(\mu) = (1-2\psi_0)^{-\frac{1}{2}(m+1)} \begin{vmatrix} \mu^m & -h_1 & \dots & -h_m \\ \mu^{m-1} & (1-2\psi_0)^{\frac{1}{2}} & \dots & -h_{m-1} \\ \mu^{m-2} & 0 & \dots & -h_{m-2} \\ \dots & \dots & \dots & \dots \\ 1 & 0 & \dots & (1-2\psi_0)^{\frac{1}{2}} \end{vmatrix}. \tag{29·42}$$

Equations (29·32), (29·41) and (29·42) give the final expression for $j_n(\mu^{-1})$ when $\psi_0 < \frac{1}{2}$. This is easily shown to agree with that found in § 29·1.

30. The solution of $(1-L)_\tau\{\mathfrak{J}(t)\} = K_n(\tau)$. (30·1)

The solution given in §§ 29·1 and 29·2 depends upon the differential difference equation

$$(d/d\tau)\,u_n = u_{n-1}, \quad u_0 = a_0$$

satisfied by the polynomial

$$u_n = \sum_{\nu=0}^{n} a_\nu \tau^{n-\nu}/(n-\nu)!\,.$$

The same method can be applied if $B(\tau) = u_n$, where

$$(d/d\tau)\,u_n = \pm\, u_{n-1}, \quad u_0 = g(\tau), \tag{30·2}$$

provided that a solution of

$$(1-L)_\tau\{\mathfrak{J}(t)\} = g(\tau) \tag{30·3}$$

can be found. In particular, the method can be used when $g(\tau) = K_1(\tau)$, $u_n = K_{n+1}(\tau)$, but in this case the solution can be found more simply. We shall prove

THEOREM 30·1. *When* $\psi_0 \leqslant \frac{1}{2}$, *the N-solution* $\Phi_n(\tau)$ *of* (30·1) *exists. It is* $C(1,1)$ *for* $n \geqslant 2$, *but* $\Phi_1(\tau) \in C(\ln \tau^{-1}, 1)$. *If*

$$\phi_n(s) = \mathfrak{L}_s\{\Phi_n(\tau)\},$$

then, for $n \geqslant 2$ *and* $\mu \geqslant 0$,

$$\phi_n(\mu^{-1}) = H(\mu) \sum_{r=0}^{n-2} h_{n-2-r}(-\mu)^r + (-1)^{n-1} \mu^{n-2} [H(\mu) - 1], \tag{30·4}$$

and for $\mu > 0$, $$\phi_1(\mu^{-1}) = [H(\mu) - 1]/\mu. \tag{30·5}$$

By Lemma 27·2, the N-solution of (30·1) exists when $n = 1$ ($\psi_0 \leqslant \frac{1}{2}$), and it is $C(\ln \tau^{-1}, 1)$. Since, for $n \geqslant 3$,

$$K_n(\tau) \leqslant K_2(\tau) \leqslant K_1(\tau),$$

therefore, for $\tau > 0$,

$$L_\tau^\nu\{K_n(t)\} \leqslant L_\tau^\nu\{K_2(t)\} \leqslant L_\tau^\nu\{K_1(t)\}. \tag{30·6}$$

Hence the N-series converges for all $n \geqslant 2$ and $\tau > 0$. Since $K_n(\tau) \geqslant 0$, the sum is the N-solution (Theorem 19·1). Let

$$\Phi_n(\tau) = \sum_{\nu=0}^{\infty} L_\tau^\nu\{K_n(t)\} \quad (n \geqslant 1, \psi_0 \leqslant \tfrac{1}{2}). \tag{30·7}$$

Then $\Phi_1(\tau) \in C(\ln \tau^{-1}, 1)$ and, by (30·6),

$$\Phi_n(\tau) \leqslant \Phi_2(\tau) \leqslant \Phi_1(\tau) \quad (n \geqslant 3). \tag{30·8}$$

It follows from Theorem 19·2 that every $\Phi_n(\tau)$ is continuous for $\tau > 0$.

When $\psi_0 = \frac{1}{2}$, $\Phi_2(\tau) = 1$ by Lemma 19·3. Hence $\Phi_n(\tau) \in C(1, 1)$ for $n \geqslant 2$.

When $\psi_0 < \frac{1}{2}$, $K_2(\tau) \leqslant \psi_0$ and therefore, by (15·6),

$$\Phi_n(\tau) \leqslant \psi_0 \sum_{\nu=0}^{\infty} (2\psi_0)^\nu = \psi_0/(1 - 2\psi_0) \quad (n \geqslant 2). \tag{30·9}$$

Hence $\Phi_n(\tau) \in C(1, 1)$ when $n \geqslant 2$.

Let $\psi_0 \leqslant \frac{1}{2}$ and let $\phi_n(s) = \mathfrak{L}_s\{\Phi_n(\tau)\}$ ($s > 0$). By (27·1), since all the functions involved are non-negative,

$$\begin{aligned} \int_1^\infty J(\tau, x) \Psi(x^{-1}) \frac{dx}{x^n} &= \sum_{\nu=0}^{\infty} L_\tau^\nu \left\{ \int_1^\infty \Psi(x^{-1}) \exp(-xt) \frac{dx}{x^n} \right\} \\ &= \sum_{\nu=0}^{\infty} L_\tau^\nu\{K_n(t)\} \\ &= \Phi_n(\tau), \end{aligned} \tag{30·10}$$

and hence, by (28·1),

$$\phi_n(s) = s\int_1^\infty R(s,x)\,\Upsilon(x^{-1})\frac{dx}{x^n}. \tag{30·11}$$

On putting $s = \mu^{-1}$, $x = u^{-1}$ and using (28·2), we get

$$\phi_n(\mu^{-1}) = H(\mu)\int_0^1 \frac{\Upsilon(u)\,H(u)}{\mu+u}\,u^{n-1}\,du. \tag{30·12}$$

But
$$\frac{u^{n-1}}{\mu+u} = \sum_{r=0}^{n-2} u^{n-2-r}(-\mu)^r + \frac{(-\mu)^{n-1}}{\mu+u}, \tag{30·13}$$

and therefore

$$\phi_n(\mu^{-1}) = H(\mu)\left\{\sum_{r=0}^{n-2}(-\mu)^r\int_0^1 u^{n-2-r}\,\Upsilon(u)\,H(u)\,du \right.$$
$$\left. +(-\mu)^{n-1}\int_0^1 \frac{\Upsilon(u)\,H(u)}{\mu+u}\,du\right\}. \tag{30·14}$$

When $n \geqslant 2$, $\mu \geqslant 0$, this reduces to (30·4) on using (8·1) and the definition of h_ν. When $n = 1$, the sum is missing in (30·13) and (30·14) and we get (30·5) when $\mu > 0$.

From Lemmas 19·2, 19·3 and 24·1, we obtain the following particular results when $\psi_0 = \frac{1}{2}$:

$$\Phi_1(\tau) = (2\psi_2)^{-\frac{1}{2}}f'(\tau), \quad \phi_1(\mu^{-1}) = [H(\mu)-1]/\mu,$$
$$\Phi_2(\tau) = 1, \quad \phi_2(\mu^{-1}) = 1,$$
$$\Phi_3(\tau) = q(\tau), \quad \phi_3(\mu^{-1}) = \sqrt{(2\psi_2)}\,H(\mu)-\mu,$$

where $f(\tau)$ and $q(\tau)$ are Hopf's functions.

The solution of

$$(1-L)_\tau\{\mathfrak{J}(t)\} = E_n(\tau) \tag{30·15}$$

can be treated similarly by omitting the function Υ in equations (30·10)–(30·12).

CHAPTER 7

FINITE ATMOSPHERES: PRELIMINARY RESULTS

31. Introduction

The fundamental equations for transfer problems in finite atmospheres were given in § 4. Such problems appear more difficult than the corresponding ones in semi-infinite atmospheres because there are two surfaces on which incident radiation may fall and through which radiation will emerge. The H-function is replaced by two functions, $X(\mu)$ and $Y(\mu)$, related by the equation

$$Y(\mu) = \exp(-\tau_1/\mu)\, X(-\mu),$$

where τ_1 is the optical depth of the atmosphere.

There are, on the other hand, simplifications. If the operator $\underline{L}$ is defined for $0 \leqslant \tau \leqslant \tau_1$ by (13·3), the homogeneous equation $(1-\underline{L})_\tau\{\mathfrak{J}(t)\} = 0$ has no non-null solution in the class of functions which we shall consider; and all proofs hold for $\psi_0 \leqslant \frac{1}{2}$.

32. Further properties of the operator $\underline{L}$

We shall be mainly concerned with two classes of functions:

The class $C(l)$ of functions which are continuous for $0 < \tau < \tau_1$ and which are $O(\ln \tau^{-1})$ as $\tau \to +0$ and $O[\ln(\tau_1-\tau)^{-1}]$ as $\tau \to \tau_1 - 0$;

The class $C(b)$ of functions which are continuous in the interval $0 \leqslant \tau \leqslant \tau_1$.†

The class $C(b)$ is contained in $C(l)$, and any function of $C(l)$ is absolutely integrable over $(0, \tau_1)$.

The deduction in Theorem 15·2 could be stated:

$$\underline{L}_\tau\{\phi(t)\} \in C(b).$$

In particular, *if* $\phi(\tau) \in C(l)$, *then* $\underline{L}_\tau\{\phi(t)\} \in C(b)$.

† See the footnote on p. 16.

If $u_n(\tau) \in C(b)$ $(n = 1, 2, \ldots)$ and if $\Sigma u_n(\tau)$ converges uniformly to $s(\tau)$ for $0 \leqslant \tau \leqslant \tau_1$, then $s(\tau)$ is continuous for $0 < \tau < \tau_1$ and tends to finite limits as $\tau \to +0$ and $\tau \to \tau_1 - 0$. Hence $s(\tau) \in C(b)$.

For $n \geqslant 2$, the functions $K_n(\tau)$ and $K_n(\tau_1 - \tau)$ belong to $C(b)$, but $K_1(\tau)$ and $K_1(\tau_1 - \tau)$ belong to $C(l)$ and not to $C(b)$.

32·1. *The integration formula.*

THEOREM 32·1. *If $\phi(\tau) \in C(l)$, then, for $0 \leqslant \tau \leqslant \tau_1$,*

$$\int_0^\tau L_t\{\phi(u)\}\,dt = L_\tau\left\{\int_0^t \phi(u)\,du\right\} - L_0\left\{\int_0^t \phi(u)\,du\right\} + [K_2(\tau_1 - \tau) - K_2(\tau_1)]\int_0^{\tau_1} \phi(u)\,du. \quad (32·1)$$

By a change of variables, we have

$$L_\tau\{\phi(t)\} = \int_0^\tau \phi(\tau - v)\,K_1(v)\,dv + \int_0^{\tau_1 - \tau} \phi(\tau + v)\,K_1(v)\,dv. \quad (32·2)$$

Hence

$$\int_0^\tau L_t\{\phi(u)\}\,dt = \int_0^\tau dt \int_0^t \phi(t - v)\,K_1(v)\,dv + \int_0^\tau dt \int_0^{\tau_1 - t} \phi(t + v)\,K_1(v)\,dv$$

$$= \int_0^\tau K_1(v)\,dv \int_v^\tau \phi(t - v)\,dt + \int_0^{\tau_1 - \tau} K_1(v)\,dv \int_0^\tau \phi(t + v)\,dt$$

$$+ \int_{\tau_1 - \tau}^{\tau_1} K_1(v)\,dv \int_0^{\tau_1 - v} \phi(t + v)\,dt,$$

the inversions being justified by absolute convergence. On changing the variable t, we get

$$\int_0^\tau L_t\{\phi(u)\}\,dt = \int_0^\tau K_1(v)\,dv \int_0^{\tau - v} \phi(u)\,du + \int_0^{\tau_1 - \tau} K_1(v)\,dv$$

$$\times \int_v^{\tau + v} \phi(u)\,du + \int_{\tau_1 - \tau}^{\tau_1} K_1(v)\,dv \int_v^{\tau_1} \phi(u)\,du.$$

Hence, by (32·2) with $\phi(t)$ replaced by $\int_0^t \phi(u)\,du$,

$$\int_0^\tau L_t\{\phi(u)\}\,dt - L_\tau\left\{\int_0^t \phi(u)\,du\right\}$$

$$= -\int_0^{\tau_1-\tau} K_1(v)\,dv \int_0^v \phi(u)\,du + \int_{\tau_1-\tau}^{\tau_1} K_1(v)\,dv \int_v^{\tau_1} \phi(u)\,du$$

$$= -\int_0^{\tau_1} K_1(v)\,dv \int_0^v \phi(u)\,du + \int_{\tau_1-\tau}^{\tau_1} K_1(v)\,dv \int_0^{\tau_1} \phi(u)\,du$$

$$= -L_0\left\{\int_0^t \phi(u)\,du\right\} + [K_2(\tau_1-\tau) - K_2(\tau_1)] \int_0^{\tau_1} \phi(u)\,du,$$

and this is (32·1).

32·2. *Preliminary lemmas.* From (32·2), we get

LEMMA 32·1.
$$L_{\tau_1-\tau}\{\phi(t)\} = L_\tau\{\phi(\tau_1 - t)\}. \tag{32·3}$$

LEMMA 32·2. *There is a constant* ρ *such that* $0 < \rho < 1$ *and*

$$0 < L_\tau\{1\} \leqslant \rho \quad (0 \leqslant \tau \leqslant \tau_1). \tag{32·4}$$

By (14·7) and (14·8),

$$L_\tau\{1\} = 2\psi_0 - K_2(\tau) - K_2(\tau_1 - \tau). \tag{32·5}$$

Hence
$$\frac{d}{d\tau} L_\tau\{1\} = K_1(\tau) - K_1(\tau_1 - \tau).$$

Since $K_1(\tau)$ is a steadily decreasing function, this changes from positive to negative as τ increases through $\frac{1}{2}\tau_1$ and hence

$$L_\tau\{1\} \leqslant 2\psi_0 - 2K_2(\tfrac{1}{2}\tau_1) = \rho,$$

and $\rho < 1$ since $K_2(\frac{1}{2}\tau_1) > 0$.

LEMMA 32·3. *There is a constant* λ *such that*

$$0 < L_\tau\{K_1(t)\} \leqslant \lambda \tag{32·6}$$

for $0 \leqslant \tau \leqslant \tau_1$.

Since $K_1(\tau) \in C(l)$, therefore $L_\tau\{K_1(t)\} \in C(b)$, and this implies (32·6) since $K_1(\tau) \geqslant 0$.

LEMMA 32·4. *For* $\nu = 1, 2, \ldots$ *and* $0 \leqslant \tau \leqslant \tau_1$

$$0 < L_\tau^\nu\{K_1(t)\} \leqslant \lambda\rho^{\nu-1}, \tag{32·7}$$

$$0 < L_\tau^\nu\{K_1(\tau_1 - t)\} \leqslant \lambda\rho^{\nu-1}. \tag{32·8}$$

The inequality (32·7) follows from (32·6) by repeated use of Lemma 32·2. Also, by Lemma 32·1,

$$\Lambda_\tau\{K_1(\tau_1 - t)\} = \Lambda_{\tau_1-\tau}\{K_1(t)\}$$

and hence (32·8) follows from (32·7).

33. The homogeneous equation

THEOREM 33·1. *If $J(\tau) \in C(l)$ and if $J(\tau)$ is a solution of*

$$(1-\Lambda)_\tau\{J(t)\} = 0, \tag{33·1}$$

then $J(\tau) = 0$.

Since $J(\tau) \in C(l)$ and since, near $\tau = 0$,

$$K_1(\tau) = \Psi(0)\ln\tau^{-1} + O(1),$$

positive constants A and B exist such that, for $0 < \tau < \tau_1$,

$$|J(\tau)| \leqslant AK_1(\tau) + BK_1(\tau_1 - \tau).$$

Hence, by Lemma 32·4,

$$|\Lambda_\tau^\nu\{J(t)\}| \leqslant A\Lambda_\tau^\nu\{K_1(t)\} + B\Lambda_\tau^\nu\{K_1(\tau_1 - t)\}$$
$$\leqslant (A+B)\,\lambda\rho^{\nu-1} \quad (\nu \geqslant 1). \tag{33·2}$$

But from (33·1),

$$J(\tau) = \Lambda_\tau\{J(t)\} = \dots = \Lambda_\tau^\nu\{J(t)\}$$

and so $$|J(\tau)| \leqslant (A+B)\,\lambda\rho^{\nu-1}.$$

Since $0 < \rho < 1$, this tends to zero as $\nu \to \infty$ and hence $J(\tau) = 0$.

34. The non-homogeneous equation

We shall make considerable use of the following theorem:

THEOREM 34·1. *Let $B(\tau) \in C(l)$. Then the N-series of the equation*

$$(1-\Lambda)_\tau\{\mathfrak{J}(t)\} = B(\tau) \tag{34·1}$$

converges to the N-solution $\mathfrak{J}_N(\tau)$ of the equation and $\mathfrak{J}_N(\tau) \in C(l)$. Also any solution of (34·1) which belongs to $C(l)$ is necessarily the N-solution. Further, $\mathfrak{J}_N(\tau_1 - \tau)$ is the N-solution of

$$(1-\Lambda)_\tau\{\mathfrak{J}(t)\} = B(\tau_1 - \tau). \tag{34·2}$$

As in the proof of Theorem 33·1, positive constants A and B exist such that, for $0<\tau<\tau_1$,

$$|B(\tau)| \leqslant AK_1(\tau)+BK_1(\tau_1-\tau), \tag{34·3}$$

and therefore (cf. (33·2))

$$|\bar{L}_\tau^\nu\{B(t)\}| \leqslant (A+B)\,\lambda\rho^{\nu-1} \quad (\nu \geqslant 1,\ 0 \leqslant \tau \leqslant \tau_1). \tag{34·4}$$

Hence the series $\sum_{\nu=1}^{\infty} \bar{L}_\tau^\nu\{B(t)\}$ is uniformly convergent for $0 \leqslant \tau \leqslant \tau_1$. Since $B(\tau)\in C(l)$, each term belongs to $C(b)$ and therefore the sum belongs to $C(b)$. It follows that, if

$$\mathfrak{J}_N(\tau) = B(\tau) + \sum_{\nu=1}^{\infty} \bar{L}_\tau^\nu\{B(t)\}, \tag{34·5}$$

then $\mathfrak{J}_N(\tau)\in C(l)$.

Let

$$\mathfrak{J}_N^n(\tau) = \sum_{\nu=0}^{n} \bar{L}_\tau^\nu\{B(t)\}. \tag{34·6}$$

Then

$$\mathfrak{J}_N^{n+1}(\tau) = \bar{L}_\tau\{\mathfrak{J}_N^n(\tau)\}+B(\tau). \tag{34·7}$$

From (34·3), (34·4) and (34·5),

$$|\mathfrak{J}_N^n(\tau)| \leqslant AK_1(\tau)+BK_1(\tau_1-\tau)+(A+B)\,\lambda/(1-\rho) = \phi(\tau).$$

Since $\bar{L}_\tau\{\phi(t)\}$ exists, it follows by dominated convergence that

$$\lim_{n\to\infty} \bar{L}_\tau\{\mathfrak{J}_N^n(t)\} = \bar{L}_\tau\{\mathfrak{J}_N(t)\},$$

and hence, from (34·7), that $\mathfrak{J}_N(\tau)$ is the N-solution. By Theorem 33·1, this is the only solution in the class $C(l)$.

Since

$$(1-\bar{L})_\tau\{\mathfrak{J}_N(t)\} = B(\tau), \tag{34·8}$$

by Lemma 32·1 (replacing τ by $\tau_1-\tau$)

$$(1-\bar{L})_\tau\{\mathfrak{J}_N(\tau_1-t)\} = B(\tau_1-\tau). \tag{34·9}$$

Hence $\mathfrak{J}_N(\tau_1-\tau)$ is a solution of (34·2), and since both $B(\tau_1-\tau)$ and $\mathfrak{J}_N(\tau_1-\tau)$ belong to $C(l)$, it must be the N-solution.

From (34·5) it is seen that, if $B(\tau)\in C(b)$, then $\mathfrak{J}_N(\tau)\in C(b)$. Hence we have

COROLLARY 1. *If* $B(\tau)\in C(b)$, *then the N-solution of* (34·1) *also belongs to* $C(b)$.

The relation between the N-solutions of (34·1) and (34·2) gives us

COROLLARY 2. *If* $B(\tau) \in C(l)$ *and if*†

$$\mathfrak{J}(\tau) = \sum_{\nu=0}^{\infty} L_{\tau}^{\nu}\{B(t)\}, \tag{34·10}$$

then

$$\mathfrak{J}(\tau_1 - \tau) = \sum_{\nu=0}^{\infty} L_{\tau}^{\nu}\{B(\tau_1 - t)\}. \tag{34·11}$$

35. The derivative of the solution of a Milne equation

The next theorem is the key to much of the analysis leading to the X- and Y-functions.

THEOREM 35·1. *Let* $B(\tau)$ *have a continuous derivative for* $0 \leqslant \tau \leqslant \tau_1$, *and let* $\mathfrak{J}(\tau)$ *denote the* N*-solution of the equation*

$$(1-L)_{\tau}\{\mathfrak{J}(t)\} = B(\tau). \tag{35·1}$$

Then $\mathfrak{J}(\tau) \in C(b)$ *and* $\mathfrak{J}'(\tau)$ *exists and belongs to* $C(l)$. *Moreover,* $\mathfrak{J}'(\tau)$ *is the* N*-solution of*

$$(1-L)_{\tau}\{\mathfrak{J}'(t)\} = B'(\tau) + \mathfrak{J}(0)\, K_1(\tau) - \mathfrak{J}(\tau_1)\, K_1(\tau_1 - \tau). \tag{35·2}$$

Since $B(\tau) \in C(b)$, therefore $\mathfrak{J}(\tau) \in C(b)$; and since the function on the right of (35·2) is $C(l)$, therefore the N-solution of (35·2) is $C(l)$. Denoting this by $G_N(\tau)$, we have

$$(1-L)_{\tau}\{G_N(t)\} = B'(\tau) + \mathfrak{J}(0)\, K_1(\tau) - \mathfrak{J}(\tau_1)\, K_1(\tau_1 - \tau). \tag{35·3}$$

Let

$$F(\tau) = \int_0^{\tau} G_N(t)\, dt - \mathfrak{J}(\tau) + \mathfrak{J}(0). \tag{35·4}$$

Then we want to prove that $F(\tau) = 0$. From this the theorem will follow. We begin by obtaining a Milne equation for $F(\tau)$.

† We now omit the subscript N, because the solution is unique.

On integrating (35·3) over $(0, \tau)$ and using Theorem 32·1, we get

$$(1-\bar{L})_\tau\left\{\int_0^t G_N(u)\,du\right\}$$
$$= B(\tau)-B(0)+\mathfrak{J}(0)\,[\psi_0-K_2(\tau)]$$
$$-\mathfrak{J}(\tau_1)\,[K_2(\tau_1-\tau)-K_2(\tau_1)]-\bar{L}_0\left\{\int_0^t G_N(u)\,du\right\}$$
$$+[K_2(\tau_1-\tau)-K_2(\tau_1)]\int_0^{\tau_1} G_N(u)\,du. \qquad (35·5)$$

From (35·1),

$$(1-\bar{L})_\tau\{\mathfrak{J}(t)\}-(1-\bar{L})_0\{\mathfrak{J}(t)\} = B(\tau)-B(0),$$

i.e. $$(1-\bar{L})_\tau\{\mathfrak{J}(t)\} = B(\tau)-B(0)+\mathfrak{J}(0)-\bar{L}_0\{\mathfrak{J}(t)\}. \qquad (35·6)$$

Also, from (32·5),

$$(1-\bar{L})_\tau\{\mathfrak{J}(0)\} = \mathfrak{J}(0)-\mathfrak{J}(0)\,[2\psi_0-K_2(\tau)-K_2(\tau_1-\tau)],$$

and $$\bar{L}_0\{\mathfrak{J}(0)\} = \mathfrak{J}(0)\,[\psi_0-K_2(\tau_1)].$$

Hence, by addition,

$$(1-\bar{L})_\tau\{\mathfrak{J}(0)\}$$
$$= \mathfrak{J}(0)\,[1-\psi_0+K_2(\tau)+K_2(\tau_1-\tau)-K_2(\tau_1)]-\bar{L}_0\{\mathfrak{J}(0)\}. \qquad (35·7)$$

From (35·4)–(35·7), we have

$$(1-\bar{L})_\tau\{F(t)\} = F(\tau_1)\,[K_2(\tau_1-\tau)-K_2(\tau_1)]-\bar{L}_0\{F(t)\}, \qquad (35·8)$$

i.e. $$(1-\bar{L})_\tau\{F(t)\} = AK_2(\tau_1-\tau)-B, \qquad (35·9)$$

where
$$\left.\begin{aligned} A &= F(\tau_1),\\ B &= F(\tau_1)\,K_2(\tau_1)+\bar{L}_0\{F(t)\}.\end{aligned}\right\} \qquad (35·10)$$

Since $G_N(\tau)\in C(l)$ and $\mathfrak{J}(\tau)\in C(b)$, it is easily verified that $F(\tau)\in C(b)$. Hence $F(\tau)$ is the N-solution of (35·9) and therefore

$$F(\tau) = A\sum_{\nu=0}^{\infty}\bar{L}_\tau^\nu\{K_2(\tau_1-t)\}-B\sum_{\nu=0}^{\infty}\bar{L}_\tau^\nu\{1\}$$
$$= AF_1(\tau)-BF_2(\tau), \qquad (35·11)$$

where $F_1(\tau)$ and $F_2(\tau)$ are positive functions which, by Corollary 1 of Theorem 34·1, belong to $C(b)$. Since $F(0) = 0$ and (by (35·10)) $F(\tau_1) = A$, therefore

$$\left.\begin{aligned} AF_1(0) - BF_2(0) &= 0, \\ A[1 - F_1(\tau_1)] + BF_2(\tau_1) &= 0. \end{aligned}\right\} \tag{35·12}$$

We shall show that $A = B = 0$ and hence, from (35·11), that $F(\tau) = 0$.

By Corollary 2 of Theorem 34·1,

$$F_2(\tau) = \sum_{\nu=0}^{\infty} L_\tau^\nu\{1\} = F_2(\tau_1 - \tau), \tag{35·13}$$

and hence $F_2(0) = F_2(\tau_1)$. Thus the equations (35·12) give

$$A[1 - F_1(\tau_1) + F_1(0)] = 0. \tag{35·14}$$

Since $L_\tau\{1\} = 2\psi_0 - K_2(\tau) - K_2(\tau_1 - \tau)$, therefore

$$L_\tau^{\nu+1}\{1\} = 2\psi_0 L_\tau^\nu\{1\} - L_\tau^\nu\{K_2(t)\} - L_\tau^\nu\{K_2(\tau_1 - t)\}. \tag{35·15}$$

Summing from $\nu = 0$ to ∞ and using the definitions of $F_1(\tau)$ and $F_2(\tau)$ (see (35·11)), we get

$$F_2(\tau) - 1 = 2\psi_0 F_2(\tau) - F_1(\tau_1 - \tau) - F_1(\tau). \tag{35·16}$$

Hence (putting $\tau = 0$)

$$1 - F_1(\tau_1) = (1 - 2\psi_0) F_2(0) + F_1(0), \tag{35·17}$$

and (35·14) can be written

$$A[(1 - 2\psi_0) F_2(0) + 2F_1(0)] = 0.$$

But $\psi_0 \leq \frac{1}{2}$, $F_2(0) > 0$, $F_1(0) > 0$, and hence $A = 0$. From (35·12) it follows that $B = 0$ also. This completes the proof.

CHAPTER 8

THE X- AND Y-FUNCTIONS

36. Introduction

The X- and Y-functions were first introduced by Chandrasekhar (see [4]) using principles of invariance, and he studied them in detail, though without rigour. We use his notation.† His introduction of the functions was so formidable that few people have ventured to work on problems involving them. Moreover, when *Radiative Transfer* was written, the existence of the functions had not been established—though they had been calculated for a limited range of values of τ_1—and some problems connected with their uniqueness were not fully understood.

The treatment given here uses the Ambartsumian technique‡ combined with the theory of N-solutions. $X(\mu)$ and $Y(\mu)$ are defined in terms of the N-solution of the auxiliary equation so that their existence is assured, and they are then shown to satisfy Chandrasekhar's X- and Y-equations. The use of N-solutions has resolved most of the problems of uniqueness.

37. The auxiliary equation

This is of the same form as before with $\bar{L}$ replacing L, viz.

$$(1-\bar{L})_\tau\{J(t,\sigma)\} = \exp(-\sigma\tau), \tag{37·1}$$

where $J(\tau,\sigma)$ denotes the N-solution, viz.

$$J(\tau,\sigma) = \sum_{\nu=0}^{\infty} \bar{L}_\tau^\nu\{\exp(-\sigma t)\}. \tag{37·2}$$

When $B(\tau) = \exp(-\sigma\tau)$, the conditions of Theorem 35·1 are satisfied for any finite value of σ. The series (37·2) converges to a function of $C(b)$ and $(\partial/\partial\tau)J(\tau,\sigma)$ exists and is $C(l)$. The

† In Russia the functions are denoted by $\phi(\mu)$ and $\psi(\mu)$ and are called 'Ambartsumian's functions'.

‡ This was done formally in Busbridge [3], Part I. Sobolev, in Russia, appears to have used an almost identical treatment for $\phi(\mu)$ and $\psi(\mu)$ at about the same time. His work is formal. See Sobolev [5], chap. VI.

latter is the N-solution of

$$(1-\bar{L})_\tau\left\{\frac{\partial}{\partial t}J(t,\sigma)\right\} = -\sigma\exp(-\sigma\tau) + J(0,\sigma)K_1(\tau) - J(\tau_1,\sigma)K_1(\tau_1-\tau), \qquad (37\cdot3)$$

and hence we have†

LEMMA 37·1. *If*

$$F(\tau) = (\partial/\partial\tau)J(\tau,\sigma) + \sigma J(\tau,\sigma), \qquad (37\cdot4)$$

then $F(\tau)$ is the N-solution of

$$(1-\bar{L})_\tau\{F(t)\} = J(0,\sigma)K_1(\tau) - J(\tau_1,\sigma)K_1(\tau_1-\tau). \qquad (37\cdot5)$$

By a proof similar to that of Lemma 27·1, we have

LEMMA 37·2. *If $0 \leqslant \tau \leqslant \tau_1$, $J(\tau,\sigma)$ is a continuous function of σ in any finite interval. If $0 < \tau \leqslant \tau_1$, $J(\tau,\sigma) \to 0$ as $\sigma \to +\infty$, but $J(0,\sigma) \to 1$ as $\sigma \to +\infty$.*

LEMMA 37·3. *If*

$$\mathfrak{J}(\tau) = \int_1^\infty J(\tau,x)\Psi(x^{-1})\frac{dx}{x}, \qquad (37\cdot6)$$

then $\mathfrak{J}(\tau)$ is the N-solution of

$$(1-\bar{L})_\tau\{\mathfrak{J}(t)\} = K_1(\tau), \qquad (37\cdot7)$$

and $\mathfrak{J}(\tau) \in C(l)$.

As in the proof of Lemma 27·2,

$$\int_1^\infty J(\tau,x)\Psi(x^{-1})\frac{dx}{x} = \sum_{\nu=0}^\infty \bar{L}_\tau^\nu\{K_1(t)\}. \qquad (37\cdot8)$$

Hence $\mathfrak{J}(\tau)$ is the N-solution of (37·7). Since $K_1(\tau) \in C(l)$, therefor $\mathfrak{J}(\tau) \in C(l)$ (Theorem 34·1).

LEMMA 37·4.

$$\int_0^{\tau_1} J(\tau,\sigma)\exp(-s\tau)\,d\tau = \int_0^{\tau_1} J(\tau,s)\exp(-\sigma\tau)\,d\tau. \qquad (37\cdot9)$$

$$\int_0^{\tau_1} J(\tau,\sigma)\exp[-s(\tau_1-\tau)]\,d\tau = \int_0^{\tau_1} J(\tau,s)\exp[-\sigma(\tau_1-\tau)]\,d\tau. \qquad (37\cdot10)$$

† This lemma embodies the first step in the Ambartsumian technique.

The first of these is obtained by applying Theorem 19·3 (with $\bar{L}$ in place of L) to (37·1) and the similar equation with s in place of σ. To prove the second, replace τ by $\tau_1-\tau$ and σ by s in (37·1). Then, by the last part of Theorem 34·1, $J(\tau_1-\tau, s)$ is the N-solution of

$$(1-\bar{L})_\tau\{J(\tau_1-t, s)\} = \exp[-s(\tau_1-\tau)]. \qquad (37\cdot11)$$

On applying Theorem 19·3 to (37·1) and (37·11), we get

$$\int_0^{\tau_1} J(\tau,\sigma)\exp[-s(\tau_1-\tau)]\,d\tau = \int_0^{\tau_1} J(\tau_1-\tau, s)\exp(-\sigma\tau)\,d\tau,$$

and on putting $t = \tau_1-\tau$ on the right, we obtain (37·10).

We can now prove (cf. Theorem 28·1)

THEOREM 37·1. *Let $J(\tau,\sigma)$ denote the N-solution of* (37·1) *and let*

$$R(s,\sigma) = \int_0^{\tau_1} J(\tau,\sigma)\exp(-s\tau)\,d\tau. \qquad (37\cdot12)$$

Then $$R(s,\sigma) = [X(s^{-1})X(\sigma^{-1}) - Y(s^{-1})Y(\sigma^{-1})]/(s+\sigma), \qquad (37\cdot13)$$

where $X(\mu)$ and $Y(\mu)$ are defined for $\mu \neq 0$ by

$$X(\mu) = J(0,\mu^{-1}), \quad Y(\mu) = J(\tau_1,\mu^{-1}). \qquad (37\cdot14)$$

The functions $X(\mu)$ and $Y(\mu)$ satisfy the equations

$$X(\mu) = 1+\mu\int_0^1\{X(\mu)X(u) - Y(\mu)Y(u)\}\frac{\Psi(u)}{\mu+u}\,du, \qquad (37\cdot15)$$

$$Y(\mu) = \exp(-\tau_1/\mu)+\mu\int_0^1\{Y(\mu)X(u) - X(\mu)Y(u)\}\frac{\Psi(u)}{\mu-u}\,du, \qquad (37\cdot16)$$

and are related by the equation

$$Y(\mu) = \exp(-\tau_1/\mu)X(-\mu). \qquad (37\cdot17)$$

From Lemma 37·3 (with τ also replaced by $\tau_1-\tau$) it follows that, if

$$F(\tau) = J(0,\sigma)\int_1^\infty J(\tau,x)\Psi(x^{-1})\frac{dx}{x} - J(\tau_1,\sigma)\int_1^\infty J(\tau_1-\tau,x)\Psi(x^{-1})\frac{dx}{x},$$

then $F(\tau)$ is the N-solution of

$$(1-\mathbf{L})_\tau\{F(t)\} = J(0,\sigma)\,K_1(\tau) - J(\tau_1,\sigma)\,K_1(\tau_1-\tau).$$

By Lemma 37·1, the N-solution of this equation is

$$(\partial/\partial\tau)\,J(\tau,\sigma) + \sigma J(\tau,\sigma),$$

and hence (cf. (28·8) in Theorem 28·1)

$$(\partial/\partial\tau)\,J(\tau,\sigma) + \sigma J(\tau,\sigma) = J(0,\sigma)\int_1^\infty J(\tau,x)\,\Psi(x^{-1})\frac{dx}{x} - J(\tau_1,\sigma)\int_1^\infty J(\tau_1-\tau,x)\,\Psi(x^{-1})\frac{dx}{x}. \tag{37·18}$$

Defining $R(s,\sigma)$ by (37·12), we have, by Lemma 37·4,†

$$R(s,\sigma) = R(\sigma,s), \tag{37·19}$$

$$\exp(-s\tau_1)\,R(-s,\sigma) = \exp(-\sigma\tau_1)\,R(-\sigma,s). \tag{37·20}$$

Multiply (37·18) by $\exp(-s\tau)$ and integrate over $(0,\tau_1)$, integrating the first term by parts. On inverting the orders of integration on the right,‡ we get

$$\begin{aligned}\exp(-s\tau_1)\,&J(\tau_1,\sigma) - J(0,\sigma) + sR(s,\sigma) + \sigma R(s,\sigma)\\ &= J(0,\sigma)\int_1^\infty \Psi(x^{-1})\frac{dx}{x}\int_0^{\tau_1} J(\tau,x)\exp(-s\tau)\,d\tau\\ &\qquad - J(\tau_1,\sigma)\int_1^\infty \Psi(x^{-1})\frac{dx}{x}\int_0^{\tau_1} J(\tau_1-\tau,x)\exp(-s\tau)\,d\tau\\ &= J(0,\sigma)\int_1^\infty \Psi(x^{-1})\,R(s,x)\frac{dx}{x}\\ &\qquad - J(\tau_1,\sigma)\exp(-s\tau_1)\int_1^\infty \Psi(x^{-1})\,R(-s,x)\frac{dx}{x}.\end{aligned}$$

† If

$$S(\tau_1,\mu,\mu_0) = R(\mu^{-1},\mu_0^{-1}) \quad \text{and} \quad T(\tau_1,\mu,\mu_0) = \exp(-\tau_1/\mu)\,R(-\mu^{-1},\mu_0^{-1}),$$

then these are scattering and transmission functions of Chandrasekhar [1]. Further relations between these functions and between the X- and Y-functions are given in Ueno [3] and Sobolev [4].

‡ This and subsequent inversions are justified by absolute convergence.

Hence

$$(s+\sigma)R(s,\sigma) = J(0,\sigma)\left\{1+\int_1^\infty \Psi(x^{-1})R(s,x)\frac{dx}{x}\right\}$$
$$-J(\tau_1,\sigma)\exp(-s\tau_1)\left\{1+\int_1^\infty \Psi(x^{-1})R(-s,x)\frac{dx}{x}\right\}. \tag{37·21}$$

Put $\tau = 0$ and write s for σ in (37·1). Then

$$J(0,s) = 1+\int_0^{\tau_1} J(t,s)K_1(t)\,dt$$
$$= 1+\int_0^{\tau_1} J(t,s)\,dt\int_1^\infty \Psi(x^{-1})\exp(-xt)\frac{dx}{x}$$
$$= 1+\int_1^\infty \Psi(x^{-1})R(x,s)\frac{dx}{x},$$

and hence, by (37·19),

$$J(0,s) = 1+\int_1^\infty \Psi(x^{-1})R(s,x)\frac{dx}{x}. \tag{37·22}$$

Similarly, putting $\tau = \tau_1$ in (37·1), we have

$$J(\tau_1,s) = \int_0^{\tau_1} J(t,s)K_1(\tau_1-t)\,dt+\exp(-s\tau_1)$$
$$= \exp(-s\tau_1)+\int_0^{\tau_1} J(t,s)\,dt\int_1^\infty \Psi(x^{-1})\exp[-x(\tau_1-t)]\frac{dx}{x}$$
$$= \exp(-s\tau_1)+\int_1^\infty \Psi(x^{-1})\exp(-x\tau_1)R(-x,s)\frac{dx}{x}$$
$$= \exp(-s\tau_1)\left\{1+\int_1^\infty \Psi(x^{-1})R(-s,x)\frac{dx}{x}\right\} \tag{37·23}$$

by (37·20).

Equations (37·21)–(37·23) give

$$(s+\sigma)R(s,\sigma) = J(0,\sigma)J(0,s)-J(\tau_1,\sigma)J(\tau_1,s), \tag{37·24}$$

and on defining $X(\mu)$ and $Y(\mu)$ by (37·14), we get (37·13). Also, from (37·22) and (37·23),

$$J(\tau_1,s) = \exp(-s\tau_1)J(0,-s) \tag{37·25}$$

and hence we have (37·17).

Finally, (37·15) follows from (37·22) and (37·24) on writing $s = 1/\mu$, $x = 1/u$ and using (37·14); while (37·16) comes similarly from (37·23)–(37·25). Alternatively, (37·16) can be deduced from (37·15) by means of the relation (37·17).

38. Properties of the X- and Y-functions

The existence of $X(\mu)$ and $Y(\mu)$ for all real μ except $\mu = 0$ is established by Theorem 37·1. The series (37·2) for $J(\tau, \sigma)$ remains absolutely convergent if σ is complex, and the equations (37·14) can therefore be used to define $X(\mu)$ and $Y(\mu)$ for complex values of μ. We now prove

THEOREM 38·1. *$X(\mu)$ and $Y(\mu)$ are analytic functions of the complex variable μ, which are regular for $|\mu| > 0$ (infinity included). Both functions are positive when μ is real ($\mu \neq 0$). Also*

$$X(\mu) \to 1 \quad as \quad |\mu| \to 0 \quad (|\arg \mu| \leqslant \tfrac{1}{2}\pi), \tag{38·1}$$

$$Y(\mu) \to 0 \quad as \quad |\mu| \to 0 \quad (|\arg \mu| \leqslant \delta < \tfrac{1}{2}\pi), \tag{38·2}$$

the convergence being uniform with respect to $\arg \mu$. *Finally,*

$$X(\mu) \to J(0,0), \quad Y(\mu) \to J(0,0) \quad as \quad \mu \to \infty. \tag{38·3}$$

If $0 \leqslant \tau \leqslant \tau_1$, we have, from (37·2),

$$\begin{aligned} J(\tau, \sigma) &= \sum_{\nu=0}^{\infty} L_\tau^\nu \left\{ \sum_{m=0}^{\infty} \frac{(-\sigma t)^m}{m!} \right\} \\ &= \sum_{m=0}^{\infty} \frac{(-\sigma)^m}{m!} \sum_{\nu=0}^{\infty} L_\tau^\nu \{t^m\} \end{aligned} \tag{38·4}$$

if the inversion of the orders of summation and of the successive integrations is justified. This will be so, by absolute convergence, if

$$\sum_{m=0}^{\infty} \sum_{\nu=0}^{\infty} \frac{|\sigma|^m}{m!} L_\tau^\nu \{t^m\}$$

is convergent. By Lemma 32·2,

$$\frac{|\sigma|^m}{m!} L_\tau^\nu \{t^m\} \leqslant \frac{|\sigma|^m \tau_1^m}{m!} L_\tau^\nu \{1\} \leqslant \frac{|\sigma|^m \tau_1^m \rho^\nu}{m!}$$

and

$$\sum_{m=0}^{\infty} \sum_{\nu=0}^{\infty} \frac{|\sigma|^m \tau_1^m \rho^\nu}{m!} = \frac{\exp(|\sigma| \tau_1)}{1-\rho},$$

so the process is justified for $|\sigma| < \infty$.

On putting $\tau = 0$ and τ_1, and $\sigma = \mu^{-1}$ in (38·4), we get

$$X(\mu) = \sum_{m=0}^{\infty} \frac{(-1)^m C_m}{m!} \mu^{-m} \quad (|\mu| > 0), \tag{38·5}$$

$$Y(\mu) = \sum_{m=0}^{\infty} \frac{(-1)^m D_m}{m!} \mu^{-m} \quad (|\mu| > 0), \tag{38·6}$$

where
$$C_m = \left[\sum_{\nu=0}^{\infty} L_\tau^\nu\{t^m\}\right]_{\tau=0}, \tag{38·7}$$

$$D_m = \left[\sum_{\nu=0}^{\infty} L_\tau^\nu\{t^m\}\right]_{\tau=\tau_1}. \tag{38·8}$$

From (38·5) and (38·6), the analytic properties of $X(\mu)$ and $Y(\mu)$ follow.

When μ is real, $J(\tau, \mu^{-1})$ is positive and therefore $X(\mu)$ and $Y(\mu)$ are positive.

Let $\mu = re^{i\theta}$ $(|\theta| \leqslant \frac{1}{2}\pi)$ and let $0 \leqslant t \leqslant \tau_1$. Then

$$|\exp(-t/\mu)| = \exp(-tr^{-1}\cos\theta) \leqslant 1,$$

and hence, by Lemma 32·2,

$$|L_\tau^\nu\{\exp(-t/\mu)\}| \leqslant L_\tau^\nu\{1\} \leqslant \rho^\nu.$$

It follows from (37·2) that

$$|J(\tau, \mu^{-1})| \leqslant (1-\rho)^{-1} \quad (0 \leqslant \tau \leqslant \tau_1),$$

and, in particular, that

$$|X(\mu)| \leqslant (1-\rho)^{-1}, \quad |Y(\mu)| \leqslant (1-\rho)^{-1}. \tag{38·9}$$

Since, for $0 \leqslant u \leqslant 1$ (μ being restricted as before)

$$|\mu + u| \geqslant \sqrt{(r^2 + u^2)},$$

it follows from (37·15) and (38·9) that

$$|X(\mu) - 1| \leqslant 2r(1-\rho)^{-2} \int_0^1 \frac{\Psi(u)}{\sqrt{(r^2+u^2)}} du$$

$$\leqslant 2(1-\rho)^{-2} \int_0^\eta \Psi(u)\, du + \frac{2r}{\eta}(1-\rho)^{-2} \int_\eta^1 \Psi(u)\, du,$$

where $0<\eta<1$. By choosing η (small) first and then taking r sufficiently small, we have

$$|X(\mu)-1| = o(1) \quad \text{as} \quad r\to 0,$$

i.e.

$$X(\mu)\to 1 \quad \text{as} \quad |\mu|\to 0,$$

and this is uniform with respect to $\theta = \arg\mu$.

If μ is real, it follows from Lemma 37·2 that

$$Y(\mu) = J(\tau_1, \mu^{-1})\to 0 \quad \text{as} \quad \mu\to +0. \tag{38·10}$$

When $\mu = r\,e^{i\theta}$, $|\theta|\leqslant\delta<\frac{1}{2}\pi$, and $0\leqslant t\leqslant\tau_1$,

$$|\exp(-t/\mu)|\leqslant\exp(-tr^{-1}\cos\delta).$$

Hence

$$|L_\tau^\nu\{\exp(-t/\mu)\}|\leqslant L_\tau^\nu\{\exp(-tr^{-1}\cos\delta)\},$$

and therefore, by (37·2),

$$|J(\tau,\mu^{-1})|\leqslant J(\tau, r^{-1}\cos\delta). \tag{38·11}$$

Putting $\tau = \tau_1$, we have

$$|Y(\mu)|\leqslant Y(r\sec\delta)\to 0 \quad \text{as} \quad r\to 0$$

by (38·10). The convergence is again uniform with respect to θ.

By definition (putting $\sigma = 0$ in (37·1)), $J(\tau, 0)$ is the N-solution of

$$(1-L)_\tau\{J(t,0)\} = 1.$$

But, by Theorem 34·1, $J(\tau_1-\tau, 0)$ is also the N-solution of this equation, and so

$$J(\tau_1-\tau, 0) = J(\tau, 0). \tag{38·12}$$

In particular

$$J(\tau_1, 0) = J(0,0),$$

i.e.

$$\lim_{\mu\to\infty} Y(\mu) = \lim_{\mu\to\infty} X(\mu) = J(0,0).$$

This completes the proof.

THEOREM 38·2. *The expansions of* $X(\mu)$ *and* $Y(\mu)$ *for* $|\mu|>0$ *are given by the equations* (38·5)–(38·8). *The constants* C_m *and* D_m *can be expressed in terms of* τ_1 *and the moments*

$$x_n = \int_0^1 \mu^n\Psi(\mu)\,X(\mu)\,d\mu, \quad y_n = \int_0^1 \mu^n\Psi(\mu)\,Y(\mu)\,d\mu \tag{38·13}$$

by means of the equations

$$C_0 = D_0 = J(0,0) = 1/(1-x_0+y_0), \tag{38·14}$$

$$\frac{C_m}{m!} = \sum_{r=0}^{m} \frac{C_r x_{m-r}}{r!} - \sum_{r=0}^{m} \frac{D_r y_{m-r}}{r!} \quad (m \geqslant 1), \tag{38·15}$$

$$D_m = \sum_{r=0}^{m} (-1)^r \binom{m}{r} C_r \tau_1^{m-r} \quad (m \geqslant 1). \tag{38·16}$$

On letting $\mu \to \infty$ in (38·5) and (38·6), and using (38·3), we get $C_0 = D_0 = J(0,0)$. If $|\mu| > 1$, (37·15) can be written

$$X(\mu) = 1 + \int_0^1 \{X(\mu)\,X(u) - Y(\mu)\,Y(u)\}\,\Psi(u) \sum_{m=0}^{\infty} (-1)^m u^m \mu^{-m}\,du.$$

By (38·9), $X(u)$ and $Y(u)$ are bounded in $(0,1)$ as is also $\Psi(u)$, and the series on the right is uniformly convergent for $0 \leqslant u \leqslant 1$. Integrating term by term, we have

$$X(\mu) = 1 + X(\mu) \sum_{m=0}^{\infty} (-1)^m x_m \mu^{-m} - Y(\mu) \sum_{m=0}^{\infty} (-1)^m y_m \mu^{-m}. \tag{38·17}$$

For $|\mu| > 1$ these series and those for $X(\mu)$ and $Y(\mu)$ are absolutely convergent. Hence, multiplying and equating coefficients, we get

$$C_0 = 1 + C_0 x_0 - D_0 y_0,$$

which proves (38·14); and the relation (38·15) comes from the coefficients of μ^{-m} $(m \geqslant 1)$.

The relation (38·16) is derived from (37·17) on expanding and equating coefficients.

We shall only need the values of C_1 and D_1 in the conservative case. After using the relations between the moments, which are given below in Theorem 38·5, the following values are found:

When $\psi_0 = \frac{1}{2}$,

$$C_0 = D_0 = 1/2y_0, \tag{38·18}$$

$$C_1 = \tfrac{1}{2}(\tau_1 C_0 - \gamma), \quad D_1 = \tfrac{1}{2}(\tau_1 C_0 + \gamma), \tag{38·19}$$

where

$$\gamma = [x_3 - y_3 + \tfrac{1}{2}\tau_1(x_2 - y_2) + \tfrac{1}{12}\tau_1^3 y_0]/\psi_2. \tag{38·20}$$

38·1. *The moments of the X- and Y-functions.* As in § 12, moments of two kinds are usually defined: those given by (38·13) and

$$\alpha_n = \int_0^1 X(\mu)\,\mu^n\,d\mu, \quad \beta_n = \int_0^1 Y(\mu)\,\mu^n\,d\mu.$$

The latter are only of importance when $\Psi(\mu) = \frac{1}{2}\omega_0$ $(0 < \omega_0 \leqslant 1)$, because they have then been listed in the tables of X- and Y-functions compiled by Chandrasekhar, Elbert and Franklin [1]. We shall work entirely with the former.

THEOREM 38·3. *The moments x_0, y_0 satisfy the relations*

$$1 - x_0 + y_0 = 1/J(0,0), \tag{38·21}$$

$$1 - x_0 - y_0 = (1 - 2\psi_0)\,J(0,0), \tag{38·22}$$

$$1 - x_0 = [1 - 2\psi_0 + y_0^2]^{\frac{1}{2}}. \tag{38·23}$$

The relation (38·21) has been proved above (see (38·14)). To prove (38·22) we shall use a method which can often be employed to establish relations between the moments.

From the definitions of x_n and $X(\mu)$,

$$x_n = \int_0^1 \Psi(\mu)\,J(0,\mu^{-1})\,\mu^n\,d\mu = \int_1^\infty \Psi(x^{-1})\,J(0,x)\,x^{-n-2}\,dx \tag{38·24}$$

and similarly for y_n. Since (cf. (30·10))

$$\int_1^\infty \Psi(x^{-1})\,J(\tau,x)\,\frac{dx}{x^{n+2}} = \sum_{\nu=0}^{\infty} L_\tau^\nu \left\{\int_1^\infty \Psi(x^{-1})\exp(-xt)\,\frac{dx}{x^{n+2}}\right\}$$

$$= \sum_{\nu=0}^{\infty} L_\tau^\nu\{K_{n+2}(t)\}, \tag{38·25}$$

therefore

$$x_n = \left[\sum_{\nu=0}^{\infty} L_\tau^\nu\{K_{n+2}(t)\}\right]_{\tau=0}. \tag{38·26}$$

Changing τ into $\tau_1 - \tau$ in (38·25), we have (by Corollary 2 of Theorem 34·1)

$$\int_1^\infty \Psi(x^{-1})\,J(\tau_1-\tau,x)\,\frac{dx}{x^{n+2}} = \sum_{\nu=0}^{\infty} L_\tau^\nu\{K_{n+2}(\tau_1-t)\}, \tag{38·27}$$

and hence

$$y_n = \left[\sum_{\nu=0}^{\infty} L_\tau^\nu\{K_{n+2}(\tau_1-t)\}\right]_{\tau=0}. \tag{38·28}$$

But, by (32·5),

$$(1-\bar{L})_{\tau}\{1\} = 1-2\psi_0+K_2(\tau)+K_2(\tau_1-\tau),$$

and 1 is the N-solution. Hence

$$1 = (1-2\psi_0)\sum_{\nu=0}^{\infty} \bar{L}_{\tau}^{\nu}\{1\}+\sum_{\nu=0}^{\infty} \bar{L}_{\tau}^{\nu}\{K_2(t)\}+\sum_{\nu=0}^{\infty} \bar{L}_{\tau}^{\nu}\{K_2(\tau_1-t)\}.$$

Putting $\tau = 0$ and using (37·2), (38·26) and (38·28), we get

$$1 = (1-2\psi_0)\,J(0,0)+x_0+y_0,$$

and this is (38·22).

From (38·21) and (38·22), we have

$$(1-x_0)^2-y_0^2 = 1-2\psi_0,$$

and $$2(1-x_0) = (1-2\psi_0)\,J(0,0)+1/J(0,0).$$

Since $\psi_0 \leqslant \frac{1}{2}$ and $J(0,0)>0$, therefore $1-x_0>0$ and

$$1-x_0 = \{1-2\psi_0+y_0^2\}^{\frac{1}{2}}.$$

Chandrasekhar (see [1], chap. VIII, theorem 2) proved this relation by the method used in the following theorem:

THEOREM 38·4. *For* $n = 0, 1, 2, \ldots,$

$$2x_{2n} = 2\psi_{2n}+\sum_{r=0}^{2n}(-1)^r\,(x_r\,x_{2n-r}-y_r\,y_{2n-r}), \tag{38·29}$$

and in particular

$$x_2(1-x_0)+y_2\,y_0 = \psi_2-\tfrac{1}{2}(x_1^2-y_1^2). \tag{38·30}$$

Multiplying (37·15) by $\Psi(\mu)\,\mu^{2n}$ and integrating over $(0,1)$, we have

$$x_{2n} = \psi_{2n}+\int_0^1\int_0^1 \Psi(\mu)\,\Psi(u)\,\{X(\mu)\,X(u)-Y(\mu)\,Y(u)\}\frac{\mu^{2n+1}}{\mu+u}\,d\mu\,du.$$

On interchanging μ and u and adding the equations, we get

$$2x_{2n} = 2\psi_{2n}+\int_0^1\int_0^1 \Psi(\mu)\,\Psi(u)\,\{X(\mu)\,X(u)-Y(\mu)\,Y(u)\}$$
$$\times\sum_{r=0}^{2n}(-1)^r\mu^{2n-r}\,u^r\,d\mu\,du,$$

and this reduces to (38·29). Equation (38·30) is the case $n = 1$.

THEOREM 38·5. *When* $\psi_0 = \frac{1}{2}$,

$$x_0 + y_0 = 1, \tag{38·31}$$

$$x_1 - y_1 = \tau_1 y_0, \tag{38·32}$$

$$y_0[x_2 + y_2 + \tfrac{1}{2}\tau_1(x_1 + y_1)] = \psi_2. \tag{38·33}$$

Equation (38·31) comes at once from (38·22). To prove (38·32), we start from the identity

$$(1 - \bar{L})_\tau \{t\} = K_3(\tau_1 - \tau) - K_3(\tau) + \tau_1 K_2(\tau_1 - \tau), \tag{38·34}$$

which is easily verified. Since τ is the N-solution of this equation,

$$\tau = \sum_{\nu=0}^{\infty} \bar{L}_\tau^\nu\{K_3(\tau_1 - t)\} - \sum_{\nu=0}^{\infty} \bar{L}_\tau^\nu\{K_3(t)\} + \tau_1 \sum_{\nu=0}^{\infty} \bar{L}_\tau^\nu\{K_2(\tau_1 - t)\}.$$

Putting $\tau = 0$ and using (38·26) and (38·28), we have

$$0 = y_1 - x_1 + \tau_1 y_0,$$

which proves (38·32). Equation (38·33) now follows from (38·30) by (38·31) and (38·32).

39. Uniqueness

Although the solution of the auxiliary equation (37·1) is unique, it does not follow that the solutions of the X- and Y-equations are necessarily unique. In the non-conservative case the question of the uniqueness has not yet been settled, but in the conservative case Chandrasekhar has proved that there is a family of solutions. (See [1], chap. VIII, theorem 5.) His proof is given in the following subsection.

39·1. *The conservative case.*

THEOREM 39·1. *When* $\psi_0 = \frac{1}{2}$, *the* X- *and* Y-*equations also have the solutions*

$$X^*(\mu) = X(\mu) + Q\mu[X(\mu) + Y(\mu)], \tag{39·1}$$

$$Y^*(\mu) = Y(\mu) - Q\mu[X(\mu) + Y(\mu)], \tag{39·2}$$

where Q *is an arbitrary constant.*

From (39·1) and (39·2), writing $X = X(\mu)$, $X_u = X(u)$, etc.,

$$X^* X_u^* - Y^* Y_u^* = XX_u - YY_u + Q(\mu + u)(X + Y)(X_u + Y_u). \tag{39·3}$$

Hence, on using (37·15),

$$1+\mu\int_0^1 \{X^*X_u^* - Y^*Y_u^*\}\frac{\Psi_u}{\mu+u}\,du$$

$$= X + Q\mu(X+Y)\int_0^1 (X_u+Y_u)\Psi_u\,du$$

$$= X + Q\mu(X+Y)(x_0+y_0)$$

$$= X^*, \tag{39·4}$$

because $x_0+y_0 = 1$. Thus X^* and Y^* satisfy (37·15). Also, by (37·17),

$$Y^*(\mu) = \exp(-\tau_1/\mu)\,X^*(-\mu) \tag{39·5}$$

and (39·4) and (39·5) together imply that X^* and Y^* satisfy (37·16).

Let x_n^* and y_n^* denote the moments of X^* and Y^*. Since

$$X^*(\mu)+Y^*(\mu) = X(\mu)+Y(\mu),$$

therefore

$$x_n^*+y_n^* = x_n+y_n, \tag{39·6}$$

and in particular

$$x_0^*+y_0^* = 1. \tag{39·7}$$

Because of the non-uniqueness of the solutions, Chandrasekhar chose as *standard solutions* those functions for which

$$x_0^* = 1, \quad y_0^* = 0.$$

These are not the functions which arise naturally. We shall denote them by $X^s(\mu)$, $Y^s(\mu)$ and their moments by x_n^s, y_n^s, so that

$$x_0^s = 1, \quad y_0^s = 0. \tag{39·8}$$

The value of Q which gives the standard solutions is $y_0/(x_1+y_1)$, and therefore

$$X^s(\mu) = X(\mu)+y_0\mu[X(\mu)+Y(\mu)]/(x_1+y_1), \tag{39·9}$$

$$Y^s(\mu) = Y(\mu)-y_0\mu[X(\mu)+Y(\mu)]/(x_1+y_1). \tag{39·10}$$

From these relations, from Theorem 38·1, and from Theorem 38·4, which (by its method of proof) is true for any solutions of the X- and Y-equations, it is easy to deduce

THEOREM 39·2. *$X^s(\mu)$ and $Y^s(\mu)$ are analytic functions of the complex variable μ, which are regular for $0<|\mu|<\infty$; $X^s(\mu)>0$*

for $\mu>0$ *and* $Y^s(\mu)>0$ *for* $\mu<0$. *Also*

$$X^s(\mu)\to 1 \quad as \quad |\mu|\to 0 \quad (|\arg\mu|\leqslant \tfrac{1}{2}\pi), \tag{39·11}$$

$$Y^s(\mu)\to 0 \quad as \quad |\mu|\to 0 \quad (|\arg\mu|\leqslant \delta<\tfrac{1}{2}\pi), \tag{39·12}$$

the convergence being uniform with respect to $\arg\mu$. *For large* μ,

$$X^s(\mu) = \mu/(x_1^s+y_1^s)+O(1), \tag{39·13}$$

$$Y^s(\mu) = -\mu/(x_1^s+y_1^s)+O(1). \tag{39·14}$$

The moments satisfy (39·8) *and the following relations*:

$$(x_1^s)^2-(y_1^s)^2 = 2\psi_2, \tag{39·15}$$

$$x_2^s+y_2^s+\tfrac{1}{2}\tau_1(x_1^s+y_1^s) = \psi_2/y_0.† \tag{39·16}$$

40. Other integral equations for $X(\mu)$ and $Y(\mu)$

If μ does not lie between -1 and 1 in the complex plane, (37·15) and (37·16) can be written

$$X\left[1-\mu\int_0^1 \Psi_u X_u \frac{du}{\mu+u}\right]+\mu Y\int_0^1 \Psi_u Y_u \frac{du}{\mu+u} = 1, \tag{40·1}$$

$$\mu X\int_0^1 \Psi_u Y_u \frac{du}{\mu-u}+Y\left[1-\mu\int_0^1 \Psi_u X_u \frac{du}{\mu-u}\right] = \exp(-\tau_1/\mu). \tag{40·2}$$

Solving for X and Y, we have

$$X\Delta = 1-\mu\int_0^1 \Psi_u X_u \frac{du}{\mu-u}-\exp(-\tau_1/\mu)\,\mu\int_0^1 \Psi_u Y_u \frac{du}{\mu+u}, \tag{40·3}$$

$$Y\Delta = \exp(-\tau_1/\mu)\left[1-\mu\int_0^1 \Psi_u X_u \frac{du}{\mu+u}\right]-\mu\int_0^1 \Psi_u Y_u \frac{du}{\mu-u}, \tag{40·4}$$

where
$$\Delta = \left(1-\mu\int_0^1 X_u \Psi_u \frac{du}{\mu+u}\right)\left(1-\mu\int_0^1 X_v \Psi_v \frac{dv}{\mu-v}\right)$$
$$-\mu^2\int_0^1 Y_u \Psi_u \frac{du}{\mu+u}\int_0^1 Y_v \Psi_v \frac{dv}{\mu-v}. \tag{40·5}$$

† The number y_0 appears in calculations involving the moments of the standard functions.

On using the identity

$$\frac{\mu}{(\mu+u)(\mu-v)} = \frac{u}{(\mu+u)(u+v)} + \frac{v}{(\mu-v)(u+v)}, \tag{40·6}$$

(40·5) becomes

$$\begin{aligned}\Delta = 1 - \mu\int_0^1 X_u \Psi_u \frac{du}{\mu+u} - \mu\int_0^1 X_v \Psi_v \frac{dv}{\mu-v} \\ + \mu\int_0^1 u\Psi_u \frac{du}{\mu+u}\int_0^1 \Psi_v (X_u X_v - Y_u Y_v)\frac{dv}{u+v} \\ + \mu\int_0^1 v\Psi_v \frac{dv}{\mu-v}\int_0^1 \Psi_u (X_v X_u - Y_v Y_u)\frac{du}{v+u}.\end{aligned} \tag{40·7}$$

Hence, by (37·15), we get

$$\begin{aligned}\Delta &= 1 - \mu\int_0^1 X_u \Psi_u \frac{du}{\mu+u} - \mu\int_0^1 X_v \Psi_v \frac{dv}{\mu-v} \\ &\quad + \mu\int_0^1 \Psi_u (X_u - 1)\frac{du}{\mu+u} + \mu\int_0^1 \Psi_v (X_v - 1)\frac{dv}{\mu-v} \\ &= 1 - \mu\int_0^1 \Psi_u \left[\frac{1}{\mu+u} + \frac{1}{\mu-u}\right] du \\ &= T(\mu),\end{aligned} \tag{40·8}$$

where $T(\mu)$ is defined by (8·4). Thus we have

THEOREM 40·1. *$X(\mu)$ and $Y(\mu)$ are solutions of the integral equations*

$$X(\mu)\,T(\mu) = 1 - \mu\int_0^1 \frac{\Psi(u)\,X(u)}{\mu-u}\,du - \mu\exp(-\tau_1/\mu)\int_0^1 \frac{\Psi(u)\,Y(u)}{\mu+u}\,du, \tag{40·9}$$

$$Y(\mu)\,T(\mu) = \exp(-\tau_1/\mu)\left[1 - \mu\int_0^1 \frac{\Psi(u)\,X(u)}{\mu+u}\,du\right] - \mu\int_0^1 \frac{\Psi(u)\,Y(u)}{\mu-u}\,du. \tag{40·10}$$

40·1. *The converse.* Is every solution of the equations (40·9) and (40·10) also a solution of (37·15) and (37·16)?

When $\psi_0 = \frac{1}{2}$, it can be shown that, if $X^*(\mu)$, $Y^*(\mu)$ are solutions of (40·9) and (40·10) which are regular for $|\mu| > 0$, $O(\mu)$ as $\mu \to \infty$, and which tend to finite limits as $|\mu| \to 0$ in the angle $|\arg \mu| \leqslant \delta < \frac{1}{2}\pi$, then $X^*(\mu)$ and $Y^*(\mu)$ are also solutions of (37·15) and (37·16).

When $\psi_0 < \frac{1}{2}$, however, the following functions can be shown to satisfy (40·9) and (40·10) but not (37·15) and (37·16):

$$X^*(\mu) = Y(\mu) + \frac{Q\mu}{1-k^2\mu^2}[X(\mu)+Y(\mu)] + \frac{1}{1-k^2\mu^2}[X(\mu)-Y(\mu)], \quad (40·11)$$

$$Y^*(\mu) = X(\mu) - \frac{Q\mu}{1-k^2\mu^2}[X(\mu)+Y(\mu)] - \frac{1}{1-k^2\mu^2}[X(\mu)-Y(\mu)], \quad (40·12)$$

where the constant Q is given by

$$Q\left(1 - \int_0^1 \Psi(u)[X(u)+Y(u)]\frac{du}{1+ku}\right) = k\left(1 - \int_0^1 \Psi(u)[X(u)-Y(u)]\frac{du}{1+ku}\right), \quad (40·13)$$

and k^{-1} is the positive zero of $T(\mu)$. (We omit the proof because of lack of space.) It should be noted that, when $\psi_0 = \frac{1}{2}$, then $k = 0$ and the functions (40·11) and (40·12) become those defined by (39·1) and (39·2). Moreover, since $x_0 + y_0 = 1$, Q (as defined by (40·13)) is indeterminate and it is therefore arbitrary.

40·2. In this chapter τ_1 has been considered as fixed. When $X(\mu)$ and $Y(\mu)$ are considered as functions of τ_1, Sobolev [4], working with $\Psi(\mu) = \frac{1}{2}\omega_0$, has shown that they satisfy another pair of integral equations. (See also Ueno [3] IX.) These are obtained by differentiating the auxiliary equation with respect to τ_1 and then applying a process similar to the Ambartsumian technique. In § 51 the corresponding integral equations are given when ω_0 varies with τ.

CHAPTER 9

FINITE ATMOSPHERES: FURTHER RESULTS

41. The τ_1-transforms

From equations (4·11) and (4·12) of the introductory section on finite atmospheres, it is seen that, corresponding to a source function $\mathfrak{J}(\tau)$, one needs to know

$$\mathfrak{j}_0(s) = s\int_0^{\tau_1} \mathfrak{J}(\tau)\exp(-s\tau)\,d\tau, \tag{41·1}$$

$$\mathfrak{j}_1(s) = s\int_0^{\tau_1} \mathfrak{J}(\tau)\exp[-s(\tau_1-\tau)]\,d\tau. \tag{41·2}$$

These will be called the τ_1-transforms of $\mathfrak{J}(\tau)$ and the corresponding lower-case letter, with subscripts 0 and 1, will always be used to denote them. Thus the τ_1-transforms of $J^n(\tau)$ will be denoted by $j_0^n(s)$ and $j_1^n(s)$.† From (41·1) and (41·2), we have the important relation

$$\mathfrak{j}_1(s) = -\exp(-s\tau_1)\,\mathfrak{j}_0(-s). \tag{41·3}$$

If $\mathfrak{J}(\tau)\in C(l)$, the τ_1-transforms exist for any finite s and both $j_0(s)/s$ and $j_1(s)/s$ are integral functions. It is easy to prove (in the usual way) that, if $0<\tau<\tau_1$,

$$\mathfrak{J}(\tau) = \lim_{\omega\to\infty}\frac{1}{2\pi i}\int_{c-i\omega}^{c+i\omega} s^{-1}\,\mathfrak{j}_0(s)\exp(s\tau)\,ds, \tag{41·4}$$

where c may be any real number. If $\tau<0$ or if $\tau>\tau_1$, the right-hand side of (41·4) is zero. These results are given in the following theorem, the proof of which is omitted:

THEOREM 41·1. *If $\mathfrak{J}(\tau)\in C(l)$ and if $\mathfrak{j}_0(s), \mathfrak{j}_1(s)$ are defined by* (41·1) *and* (41·2), *then $s^{-1}\mathfrak{j}_0(s)$ and $s^{-1}\mathfrak{j}_1(s)$ are integral functions, and $\mathfrak{J}(\tau)$ is given by* (41·4) *for $0<\tau<\tau_1$.*

† In this chapter, because of the subscripts 0, 1, we shall write $J^n(\tau)$ in place of the usual $J_n(\tau)$.

42. The solution of

$$(1-\mathsf{L})_\tau\{\mathfrak{J}(t)\} = \sum_{r=0}^{n} a_r\,\tau^{n-r}/(n-r)!. \qquad (42\cdot1)$$

Let $J^n(\tau)$ denote the N-solution of this equation. With $B(\tau)$ equal to the polynomial on the right of (42·1), the conditions of Theorem 35·1 are satisfied. Hence $J^n(\tau)\in C(b)$, $(d/d\tau)\,J^n(\tau)$ exists, belongs to $C(l)$ and is the N-solution of

$$(1-\mathsf{L})_\tau\left\{\frac{d}{dt}J^n(t)\right\} = \sum_{r=0}^{n-1} a_r\,\tau^{n-1-r}/(n-1-r)!$$
$$+J^n(0)\,K_1(\tau)-J^n(\tau_1)\,K_1(\tau_1-\tau). \qquad (42\cdot2)$$

Hence if
$$F(\tau) = (d/d\tau)\,J^n(\tau)-J^{n-1}(\tau), \qquad (42\cdot3)$$

then $F(\tau)$ is the N-solution of

$$(1-\mathsf{L})_\tau\{F(t)\} = J^n(0)\,K_1(\tau)-J^n(\tau_1)\,K_1(\tau_1-\tau). \qquad (42\cdot4)$$

From Lemma 37·3 (with τ also replaced by $\tau_1-\tau$), it follows that

$$F(\tau) = J^n(0)\int_1^\infty J(\tau,x)\,\Psi(x^{-1})\,\frac{dx}{x}-J^n(\tau_1)\int_1^\infty J(\tau_1-\tau,x)\,\Psi(x^{-1})\,\frac{dx}{x},$$

and hence that

$$\frac{d}{d\tau}J^n(\tau)-J^{n-1}(\tau) = J^n(0)\int_1^\infty J(\tau,x)\,\Psi(x^{-1})\,\frac{dx}{x}$$
$$-J^n(\tau_1)\int_1^\infty J(\tau_1-\tau,x)\,\Psi(x^{-1})\,\frac{dx}{x}. \qquad (42\cdot5)$$

This holds for $n = 1, 2, \ldots$, and it will also hold for $n = 0$ if we define $J^{-1}(\tau)$ to be zero.

Multiply (42·5) by $\exp(-s\tau)$ and integrate with respect to τ over $(0,\tau_1)$, integrating the first term by parts. Then on introducing the τ_1-transforms, we have

$$\exp(-s\tau_1)\,J^n(\tau_1)-J^n(0)+j_0^n(s)-s^{-1}j_0^{n-1}(s)$$
$$= J^n(0)\int_1^\infty R(s,x)\,\Psi(x^{-1})\,\frac{dx}{x}$$
$$-J^n(\tau_1)\exp(-s\tau_1)\int_1^\infty R(-s,x)\,\Psi(x^{-1})\,\frac{dx}{x},$$

where $R(s,x)$ is defined by (37·12). By (37·22) and (37·23), this can be written

$$j_0^n(s) - s^{-1} j_0^{n-1}(s) = J^n(0)\, J(0,s) - J^n(\tau_1)\, J(\tau_1, s), \qquad (42\cdot6)$$

and hence, by (37·14),

$$j_0^n(\mu^{-1}) - \mu j_0^{n-1}(\mu^{-1}) = J^n(0)\, X(\mu) - J^n(\tau_1)\, Y(\mu). \qquad (42\cdot7)$$

Since $j_0^{-1}(s) = 0$, on solving we get

$$j_0^n(\mu^{-1}) = X(\mu) \sum_{r=0}^{n} J^r(0)\, \mu^{n-r} - Y(\mu) \sum_{r=0}^{n} J^r(\tau_1)\, \mu^{n-r}. \qquad (42\cdot8)$$

Also, writing $-\mu$ for μ and using (41·3) and (37·17), we have

$$j_1^n(\mu^{-1}) = X(\mu) \sum_{r=0}^{n} J^r(\tau_1)\, (-\mu)^{n-r} - Y(\mu) \sum_{r=0}^{n} J^r(0)\, (-\mu)^{n-r}. \qquad (42\cdot9)$$

The method of § 29·1 is used to find equations for $J^r(0)$ and $J^r(\tau_1)$ $(r = 0, 1, \ldots)$. Put $\tau = 0$ in (42·1); then corresponding to (29·12), we get

$$J^n(0) = \int_0^{\tau_1} J^n(t)\, K_1(t)\, dt + a_n \quad (n = 0, 1, \ldots),$$

and by similar analysis (see (29·13)), we obtain

$$J^n(0) = \int_0^1 \Psi(\mu)\, j_0^n(\mu^{-1})\, d\mu + a_n. \qquad (42\cdot10)$$

Also, putting $\tau = \tau_1$ in (42·1) and denoting the polynomial by $B(\tau)$,

$$\begin{aligned} J^n(\tau_1) &= \int_0^{\tau_1} J^n(t)\, K_1(\tau_1 - t)\, dt + B(\tau_1) \\ &= \int_0^{\tau_1} J^n(t)\, dt \int_1^{\infty} \Psi(x^{-1}) \exp[-x(\tau_1 - t)] \frac{dx}{x} + B(\tau_1) \\ &= \int_1^{\infty} \Psi(x^{-1})\, j_1^n(x) \frac{dx}{x^2} + B(\tau_1), \end{aligned}$$

and hence

$$J^n(\tau_1) = \int_0^1 \Psi(\mu)\, j_1^n(\mu^{-1})\, d\mu + \sum_{r=0}^{n} a_r \tau_1^{n-r}/(n-r)!. \qquad (42\cdot11)$$

On substituting from (42·8) and (42·9) into (42·10) and (42·11), we obtain the equations

$$J^n(0) - \sum_{r=0}^{n} J^r(0)\, x_{n-r} + \sum_{r=0}^{n} J^r(\tau_1)\, y_{n-r} = a_n, \qquad (42\cdot12)$$

$$J^n(\tau_1) - \sum_{r=0}^{n} J^r(\tau_1)\,(-1)^{n-r} x_{n-r} + \sum_{r=0}^{n} J^r(0)\,(-1)^{n-r} y_{n-r}$$

$$= \sum_{r=0}^{n} a_r \tau_1^{n-r}/(n-r)!. \qquad (42\cdot13)$$

When $\psi_0 \neq \frac{1}{2}$, the constants $J^0(0)$, $J^0(\tau_1)$, etc., can be evaluated in turn. (See § 42·1 for the cases $n = 0, 1$.) When $\psi_0 = \frac{1}{2}$, the equations (42·12) and (42·13) are found to be identical, at least for $n = 0, 1$, and another method has to be employed to get a second set of equations. This is done in § 42·2.

42·1. *The case* $\psi_0 < \frac{1}{2}$ $(n = 0, 1)$. When $n = 0$, (42·12) and (42·13) give

$$\left.\begin{aligned} J^0(0)\,(1-x_0) + J^0(\tau_1)\, y_0 &= a_0, \\ J^0(0)\, y_0 + J^0(\tau_1)\,(1-x_0) &= a_0. \end{aligned}\right\} \qquad (42\cdot14)$$

Since, by (38·23),

$$(1-x_0)^2 - y_0^2 = 1 - 2\psi_0, \qquad (42\cdot15)$$

therefore $\quad J^0(0) = J^0(\tau_1) = a_0(1 - x_0 - y_0)/(1 - 2\psi_0), \qquad (42\cdot16)$

and hence, from (42·8) and (42·9),

$$j_0^0(\mu^{-1}) = j_1^0(\mu^{-1}) = \frac{a_0(1-x_0-y_0)}{1-2\psi_0}\{X(\mu) - Y(\mu)\}. \qquad (42\cdot17)$$

When $n = 1$, (42·12) and (42·13) give

$$J^1(0)\,(1-x_0) + J^1(\tau_1)\, y_0 - J^0(0)\, x_1 + J^0(\tau_1)\, y_1 = a_1,$$

$$J^1(0)\, y_0 + J^1(\tau_1)\,(1-x_0) - J^0(0)\, y_1 + J^0(\tau_1)\, x_1 = a_0 \tau_1 + a_1.$$

From these equations, (42·15) and (42·16), we get

$$\left.\begin{aligned} (1-2\psi_0)\, J^1(0) &= a_1(1-x_0-y_0) + a_0(x_1 - y_1 - \tau_1 y_0), \\ (1-2\psi_0)\, J^1(\tau_1) &= a_1(1-x_0-y_0) - a_0[x_1 - y_1 - \tau_1(1-x_0)]. \end{aligned}\right\} \qquad (42\cdot18)$$

Hence (42·8) and (42·9) give

$$j_0^1(\mu^{-1})$$
$$= \frac{X(\mu)}{1-2\psi_0}\{(a_0\mu+a_1)(1-x_0-y_0)+a_0(x_1-y_1-\tau_1 y_0)\}$$
$$-\frac{Y(\mu)}{1-2\psi_0}\{(a_0\mu+a_1)(1-x_0-y_0)-a_0[x_1-y_1-\tau_1(1-x_0)]\},$$
$$\tag{42·19}$$
$$j_1^1(\mu^{-1})$$
$$= \frac{X(\mu)}{1-2\psi_0}\{(-a_0\mu+a_1)(1-x_0-y_0)-a_0[x_1-y_1-\tau_1(1-x_0)]\}$$
$$-\frac{Y(\mu)}{1-2\psi_0}\{(-a_0\mu+a_1)(1-x_0-y_0)+a_0(x_1-y_1-\tau_1 y_0)\}. \tag{42·20}$$

Thus we have proved

THEOREM 42·1. *If $J^1(\tau)$ is the N-solution of*

$$(1-L)_\tau\{J^1(t)\} = a_0\tau+a_1 \tag{42·21}$$

when $\psi_0<\frac{1}{2}$, then $J^1(\tau)\in C(b)$ and its τ_1-transforms are given by (42·19) *and* (42·20).

These results were first found by Horak and Lundquist [1] using 'principles of invariance'.

42·2. *The case $\psi_0=\frac{1}{2}$ $(n=0,1)$.* Since $x_0+y_0=1$, the equations (42·14) are identical and we only have one equation, viz.

$$y_0\{J^0(0)+J^0(\tau_1)\} = a_0. \tag{42·22}$$

In this case there are three possible methods of procedure. In the first, introduced by Chandrasekhar, an appeal is made to the flux integral and to the K-integral (see §§ 6 and 7). A simple way of applying this method is given in Busbridge [3], Part I, equations (8·13)–(8·18).

In the second method, which can also be employed when $\psi_0<\frac{1}{2}$, use is made of the fact that $s^{-1}j_0^n(s)$ is an integral function (Theorem 41·1). On putting $\mu=s^{-1}$ in (42·8) and using the expansions (38·5) and (38·6) for $X(\mu)$ and $Y(\mu)$, we get

$$s^{-1}j_0^n(s) = \sum_{m=0}^{\infty}\frac{(-1)^m C_m s^m}{m!}\sum_{r=0}^{n}J^r(0)\,s^{-(n-r+1)}$$
$$-\sum_{m=0}^{\infty}\frac{(-1)^m D_m s^m}{m!}\sum_{=0}^{n}J^r(\tau_1)\,s^{-(n-r+1)}.$$

On multiplying, this can contain no negative powers of s. Hence, for $r = 0, 1, \ldots, n$,

$$\sum_{m=0}^{r} \frac{(-1)^m C_m}{m!} J^{r-m}(0) = \sum_{m=0}^{r} \frac{(-1)^m D_m}{m!} J^{r-m}(\tau_1). \quad (42{\cdot}23)$$

When $n = 0$ there is one equation, viz.

$$C_0 J^0(0) = D_0 J^0(\tau_1),$$

and since (by (38·14)) $C_0 = D_0$, this with (42·22) gives

$$J^0(0) = J^0(\tau_1) = a_0/2y_0. \quad (42{\cdot}24)$$

Hence, by (42·8) and (42·9),

$$j_0^0(\mu^{-1}) = j_1^0(\mu^{-1}) = \tfrac{1}{2}a_0 y_0^{-1}\{X(\mu) - Y(\mu)\}. \quad (42{\cdot}25)$$

The third and simplest method (at least when $n = 0, 1$) is to find the limiting forms, as $\psi_0 \to \frac{1}{2}$, of the τ_1-transforms found when $\psi_0 < \frac{1}{2}$. The coefficient $(1-x_0-y_0)/(1-2\psi_0)$ in (42·17) is indeterminate when $\psi_0 = \frac{1}{2}$ but, by (42·15),

$$\frac{1-x_0-y_0}{1-2\psi_0} = \frac{1}{1-x_0+y_0} \to \frac{1}{2y_0} \quad \text{as} \quad \psi_0 \to \tfrac{1}{2}. \quad (42{\cdot}26)$$

Hence the limiting form of (42·17) is (42·25).

The third method alone will be used when $n = 1$. In order to find the limiting forms of (42·19) and (42·20), in addition to (42·26) we require

$$\lim_{\psi_0 \to \frac{1}{2}} (x_1 - y_1 - \tau_1 y_0)/(1-2\psi_0).$$

From (38·14)–(38·16) with $m = 1$, we get

$$C_1 = \frac{x_1 - y_1 - \tau_1 y_0}{(1-x_0-y_0)(1-x_0+y_0)} = \frac{x_1 - y_1 - \tau_1 y_0}{1-2\psi_0}.$$

Since C_1 is given by the same formula, viz. (38·7), for both $\psi_0 < \frac{1}{2}$ and $\psi_0 = \frac{1}{2}$,

$$\lim_{\psi_0 \to \frac{1}{2}} \frac{x_1 - y_1 - \tau_1 y_0}{1-2\psi_0} = [C_1]_{\psi_0 = \frac{1}{2}} = \frac{1}{2y_0}(\tfrac{1}{2}\tau_1 - y_0\gamma) \quad (42{\cdot}27)$$

by (38·19) and (38·18). Hence, by (42·26),

$$\lim_{\psi_0 \to \frac{1}{2}} \frac{x_1 - y_1 - \tau_1(1-x_0)}{1-2\psi_0} = -\frac{1}{2y_0}(\tfrac{1}{2}\tau_1 + y_0\gamma). \quad (42{\cdot}28)$$

Thus (42·19) and (42·20) become

$$j_0^1(\mu^{-1}) = \frac{X(\mu)}{2y_0}\{(a_0\mu+a_1)+a_0(\tfrac{1}{2}\tau_1-y_0\gamma)\}$$
$$-\frac{Y(\mu)}{2y_0}\{(a_0\mu+a_1)+a_0(\tfrac{1}{2}\tau_1+y_0\gamma)\}, \qquad (42·29)$$

$$j_1^1(\mu^{-1}) = \frac{X(\mu)}{2y_0}\{(-a_0\mu+a_1)+a_0(\tfrac{1}{2}\tau_1+y_0\gamma)\}$$
$$-\frac{Y(\mu)}{2y_0}\{(-a_0\mu+a_1)+a_0(\tfrac{1}{2}\tau_1-y_0\gamma)\}, \qquad (42·30)$$

where γ is given by (38·20). Hence we have

THEOREM 42·2. *If $J^1(\tau)$ is the N-solution of* (42·21) *when $\psi_0=\frac{1}{2}$, then $J^1(\tau)\in C(b)$ and its τ_1-transforms are given by* (42·29) *and* (42·30).

It must be emphasized that the X- and Y-functions appearing in these formulae are the ones defined by (37·14) and not the standard functions. It is the latter which have been tabulated by Chandrasekhar, Elbert and Franklin [1]. Expressed in terms of $X^s(\mu)$, $Y^s(\mu)$ and their moments, the following values are found (after a lengthy reduction) for the case $\Psi(\mu)=\frac{1}{2}$:

$$j_0^1(\mu^{-1}) = 3\mu X^s(\mu)\{a_0(x_2^s+y_2^s+\tau_1 y_1^s)-(a_0\mu+a_1)(x_1^s-y_1^s)\}$$
$$-3\mu Y^s(\mu)\{a_0(x_2^s+y_2^s+\tau_1 x_1^s)+(a_0\mu+a_1)(x_1^s-y_1^s)\}$$
$$+AX^s(\mu)-BY^s(\mu), \qquad (42·31)$$

where

$$A = a_0\{3(x_1^s-y_1^s)[x_4^s+y_4^s+\tfrac{1}{2}\tau_1(x_3^s+y_3^s)]-\tfrac{1}{24}\tau_1^3\}/\Delta$$
$$+3a_0[\tau_1 y_2^s+\tfrac{1}{4}\tau_1^2(x_1^s+y_1^s)-x_3^s+y_3^s]+3a_1\Delta, \qquad (42·32)$$

$$B = -a_0\{3(x_1^s-y_1^s)[x_4^s+y_4^s+\tfrac{1}{2}\tau_1(x_3^s+y_3^s)]-\tfrac{1}{24}\tau_1^3\}/\Delta$$
$$+3a_0[\tau_1 x_2^s+\tfrac{1}{4}\tau_1^2(x_1^s+y_1^s)+x_3^s-y_3^s]+3a_1\Delta, \qquad (42·33)$$

and

$$\Delta = x_2^s+y_2^s+\tfrac{1}{2}\tau_1(x_1^s+y_1^s). \qquad (42·34)$$

The value of $j_1^1(\mu^{-1})$ can be written down from $j_0^1(\mu^{-1})$ by means of the relations (41·3) and (37·17).

43. The solution of $(1-\underline{L})_\tau\{\mathfrak{J}(t)\} = K_n(\tau)$. (43·1)

By means of (38·25) (cf. also § 30), it is easy to prove

THEOREM 43·1. *If* $\Phi^n(\tau)$ *is the* N*-solution of* (43·1), *then* $\Phi^n(\tau)\in C(b)$ *when* $n\geqslant 2$, *and* $\Phi^1(\tau)\in C(l)$. *For* $n\geqslant 2$ *and* $\mu\geqslant 0$ *the* τ_1*-transforms are*

$$\phi_0^n(\mu^{-1}) = X(\mu)\sum_{r=0}^{n-2}(-\mu)^r x_{n-2-r} - Y(\mu)\sum_{r=0}^{n-2}(-\mu)^r y_{n-2-r} + (-1)^{n-1}\mu^{n-2}[X(\mu)-1], \tag{43·2}$$

$$\phi_1^n(\mu^{-1}) = X(\mu)\sum_{r=0}^{n-2}\mu^r y_{n-2-r} - Y(\mu)\sum_{r=0}^{n-2}\mu^r x_{n-2-r} + \mu^{n-2}[Y(\mu)-\exp(-\tau_1/\mu)]. \tag{43·3}$$

When $n=1$ *and* $\mu>0$,

$$\phi_0^1(\mu^{-1}) = [X(\mu)-1]/\mu, \tag{43·4}$$

$$\phi_1^1(\mu^{-1}) = [Y(\mu)-\exp(-\tau_1/\mu)]/\mu. \tag{43·5}$$

Since $K_n(\tau)\in C(b)$ for $n\geqslant 2$ and $K_1(\tau)\in C(l)$, the same is true of $\Phi_n(\tau)$ by Theorem 34·1 and its first corollary. From (38·25)

$$\Phi^n(\tau) = \int_1^\infty \Psi(x^{-1})\,J(\tau,x)\,x^{-n}\,dx \tag{43·6}$$

and therefore

$$\begin{aligned}\phi_0^n(s) &= s\int_0^{\tau_1}\exp(-s\tau)\,d\tau\int_1^\infty \Psi(x^{-1})\,J(\tau,x)\,x^{-n}\,dx\\ &= s\int_1^\infty \Psi(x^{-1})\,R(s,x)\,x^{-n}\,dx \quad \text{[by (37·12)]}\\ &= s\int_0^1 \Psi(u)\,R(s,u^{-1})\,u^{n-2}\,du.\end{aligned}$$

On substituting for $R(s,u^{-1})$ from (37·13), we get

$$\phi_0^n(\mu^{-1}) = \int_0^1 \Psi(u)\{X(\mu)\,X(u) - Y(\mu)\,Y(u)\}\frac{u^{n-1}}{\mu+u}\,du. \tag{43·7}$$

By the identity (30·13) (the sum being absent when $n=1$),

$$\begin{aligned}\phi_0^n(\mu^{-1}) = {}& X(\mu)\sum_{r=0}^{n-2}(-\mu)^r\int_0^1 \Psi(u)\,X(u)\,u^{n-2-r}\,du\\ &- Y(\mu)\sum_{r=0}^{n-2}(-\mu)^r\int_0^1 \Psi(u)\,Y(u)\,u^{n-2-r}\,du\\ &+(-\mu)^{n-1}\int_0^1 \Psi(u)\{X(\mu)\,X(u)-Y(\mu)\,Y(u)\}\frac{du}{\mu+u},\end{aligned}$$

and on using (37·15), we get (43·2) when $n \geqslant 2$ and (43·4) when $n = 1$.

Equations (43·3) and (43·5) follow from (43·2) and (43·4) by the relations (41·3) and (37·17).

When $\psi_0 = \frac{1}{2}$, since $x_0 + y_0 = 1$, we get

$$\phi_0^2(\mu^{-1}) = 1 - y_0[X(\mu) + Y(\mu)], \tag{43·8}$$

$$\phi_1^2(\mu^{-1}) = y_0[X(\mu) + Y(\mu)] - \exp(-\tau_1/\mu). \tag{43·9}$$

Expressed in terms of the standard functions, these become [on using (39·16)]

$$\phi_0^2(\mu^{-1}) = 1 - \psi_2[X^s(\mu) + Y^s(\mu)]/\Delta, \tag{43·10}$$

$$\phi_1^2(\mu^{-1}) = \psi_2[X^s(\mu) + Y^s(\mu)]/\Delta - \exp(-\tau_1/\mu), \tag{43·11}$$

where Δ is given by (42·34). These formulae are likely to be useful in practical applications.

CHAPTER 10

ANISOTROPIC SCATTERING

44. Introduction

Although scattering, in practical cases, is usually anisotropic, little rigorous work has been done on the subject and even the formal treatment is incomplete. So far there have been two main methods of attack (apart from computational ones): that used by Chandrasekhar in [1] and that developed independently by Davison [1] and Kuščer [1] in connexion with neutron-transport theory.

Chandrasekhar's method employs 'principles of invariance'. Equations for anisotropic scattering are derived and these are then reduced to H-equations (if the atmosphere is semi-infinite) by an inspired insight denied to the reader.

The Davison–Kuščer method has been mainly used when there is axial symmetry and the medium is non-emitting. If the phase function $p(\mu, \mu')$ is given by (2·7) (see (45·1) below), the source function can be expressed in the form

$$\mathfrak{J}(\tau, \mu) = \frac{1}{2} \sum_{\nu=0}^{N} \omega_\nu P_\nu(\mu) G_\nu(\tau).$$

The Milne equation for $\mathfrak{J}(\tau, \mu)$ gives rise to N simultaneous integral equations for the $G_\nu(\tau)$, and these can be reduced to a non-homogeneous Milne equation (of the type studied in Chapter 6) for $G_0(\tau)$ by means of a differential-difference equation for the $G_\nu(\tau)$ deduced from the equation of transfer. When $G_0(\tau)$ has been found, the other $G_\nu(\tau)$ can be evaluated. Kuščer uses the Ambartsumian technique to solve the equation for $G_0(\tau)$, while Davison employs a modified Wiener–Hopf method.

For the case of axial symmetry with a phase function of the form (45·1), the Davison–Kuščer method is satisfactory, and Kuščer has worked out many details of the analysis. If an expression is wanted for $\mathfrak{J}(\tau, \mu)$, this method is better than that described below, for the latter only produces the emergent intensity $\mathfrak{j}(\mu)$, and it is no longer true that $\mathfrak{J}(\tau,\mu) = \mathfrak{L}_\tau^{-1}\{\mathfrak{j}(\mu)\}$. The Davison–Kuščer method has, however, disadvantages from a theoretical point of view: Extensions of the method to (say) scattering according to a phase matrix (see § 50) are not obvious; the formulae lack symmetry when there should be symmetry; the work does not tie up in any way with that of Chandrasekhar; the homogeneous Milne equation for $\mathfrak{J}(\tau,\mu)$ does not give rise to a homogeneous equation for $G_0(\tau)$ and the existence of a non-null solution $\mathfrak{J}(\tau,\mu)$ cannot therefore be deduced from Theorems 22·1 and 22·2.

For all these reasons a different treatment is presented here. Only the axially symmetric case of a semi-infinite atmosphere is considered, but other cases can be treated similarly. Some simple cases of the homogeneous equation are dealt with first (§ 46). These are the ones which are most likely to arise in practical work, and are those which would be needed in a rigorous treatment of the subsequent sections.

In § 47 the Ambartsumian technique is used to derive equations for the scattering function from the appropriate auxiliary equation. These equations are identical with those found by Chandrasekhar from principles of invariance, and in § 48 they are shown to have solutions in terms of H-functions. This section is long, but the results obtained are general and they include those found by Chandrasekhar. Because of the complexity of the formal analysis, details of rigour are omitted from §§ 47 and 48. It is not difficult to justify all the analysis when $0 < \omega_0 < 1$, but the conservative case presents difficulties which have not been entirely overcome.

A complete treatment of anisotropic scattering would require another book, and a selection of topics has had to be made. Following on the treatment given here, solutions of the homogeneous equation (under more general conditions) can be considered by Ambartsumian's original method (see [1]). Nonhomogeneous equations of certain types can be treated by an extension of the method of Chapter 6. Alternatively, the emergent intensity can often be found by means of the 'law of diffuse reflexion', which is established in § 49.

45. Preliminary formulae

When there is axial symmetry, the phase function of § 2 is given by

$$p(\mu,\mu') = \sum_{\nu=0}^{N} \omega_\nu P_\nu(\mu) P_\nu(\mu'), \tag{45·1}$$

where $-1 \leqslant \mu \leqslant 1$, $-1 \leqslant \mu' \leqslant 1$. It is such that (see (2·3), (2·5), (2·9)),

$$p(\mu,\mu') > 0, \tag{45·2}$$

$$p(\mu,\mu') = p(\mu',\mu), \quad p(-\mu,\mu') = p(\mu,-\mu'), \tag{45·3}$$

$$\frac{1}{2}\int_{-1}^{1} p(\mu,\mu')\,d\mu' = \omega_0 \quad (0 < \omega_0 \leqslant 1), \tag{45·4}$$

$$\frac{1}{2}\int_{-1}^{1} \mu' p(\mu,\mu')\,d\mu' = \tfrac{1}{3}\omega_1 \mu. \tag{45·5}$$

We shall assume that

$$|\omega_n| < (2n+1)\,\omega_0 \quad (n = 1, 2, \ldots, N). \tag{45·6}$$

The linear operator $\Lambda_{\tau,\mu}$ will be defined by

$$\Lambda_{\tau,\mu}\{\phi(t,\mu')\} = \frac{1}{2}\int_0^1 p(\mu,\mu')\frac{d\mu'}{\mu'}\int_\tau^\infty \phi(t,\mu')\exp[-(t-\tau)/\mu']\,dt$$
$$+\frac{1}{2}\int_0^1 p(\mu,-\mu')\frac{d\mu'}{\mu'}\int_0^\tau \phi(t,-\mu')\exp[-(\tau-t)/\mu']\,dt. \tag{45·7}$$

If

$$k(t,\mu) = \begin{cases} \frac{1}{2}|\mu|^{-1}\exp(-t/\mu) & (t/\mu > 0), \\ 0 & (t/\mu < 0), \end{cases} \tag{45·8}$$

then

$$k(t,\mu) = k(-t,-\mu), \tag{45·9}$$

and (45·7) can be written

$$\Lambda_{\tau,\mu}\{\phi(t,\mu')\} = \int_{-1}^{1} p(\mu,\mu')\,d\mu'\int_0^\infty \phi(t,\mu')\,k(t-\tau,\mu')\,dt. \tag{45·10}$$

With this notation, the symmetrical property of L, which is expressed by (15·23), is replaced by

$$\int_0^\infty \phi(\tau)\,\Lambda_{\tau,\mu}\{p(\mu',-\mu_0)\,\chi(t)\}\,d\tau = \int_0^\infty \chi(\tau)\,\Lambda_{\tau,\mu_0}\{p(\mu',-\mu)\,\phi(t)\}\,d\tau, \tag{45·11}$$

where each side of the equation is a function of μ and μ_0. For the left-hand side is

$$\int_0^\infty \phi(\tau)\,d\tau \int_{-1}^{1} p(\mu,\mu')\,d\mu' \int_0^\infty p(\mu',-\mu_0)\,\chi(t)\,k(t-\tau,\mu')\,dt$$

$$= \int_0^\infty \chi(t)\,dt \int_{-1}^{1} p(\mu',-\mu_0)\,d\mu' \int_0^\infty p(\mu,\mu')\,\phi(\tau)\,k(t-\tau,\mu')\,d\tau.$$

On writing $-\mu'$ for μ', interchanging t and τ, and using (45·3) and (45·9), we get

$$\int_0^\infty \chi(\tau)\,d\tau \int_{-1}^{1} p(\mu_0,\mu')\,d\mu' \int_0^\infty p(\mu',-\mu)\,\phi(t)\,k(t-\tau,\mu')\,dt,$$

and this is the right-hand side of (45·11). The analysis will hold when all the integrals are absolutely convergent.

The differentiation formula (cf. (15·17)) is

$$\frac{\partial}{\partial\tau}\Lambda_{\tau,\mu}\{\phi(t,\mu')\} = \Lambda_{\tau,\mu}\left\{\frac{\partial}{\partial t}\phi(t,\mu')\right\} + \frac{1}{2}\int_0^1 p(\mu,-u)\,\phi(0,-u)\exp(-\tau/u)\frac{du}{u}. \tag{45·12}$$

This will hold if $\phi(\tau,\mu)$ is a continuous function of τ and μ for $\tau \geqslant 0$, $-1 \leqslant \mu \leqslant 1$, and if it has a continuous τ-derivative for $\tau > 0$, $-1 \leqslant \mu \leqslant 1$, which is $O(\ln\tau^{-1})$ as $\tau \to +0$ and $O[\exp(a\tau)]$ as $\tau \to \infty$ $(0 \leqslant a < 1)$ uniformly for $-1 \leqslant \mu \leqslant 1$.

46. The homogeneous equation

We shall only consider solutions of

$$J(\tau,\mu) = \Lambda_{\tau,\mu}\{J(t,\mu')\} \tag{46·1}$$

which are at most $O(\tau)$ as $\tau \to \infty$.

46·1. *The non-conservative case.*

THEOREM 46·1. *Let* $0 < \omega_0 < 1$ *and let* $J(\tau,\mu)$ *be a solution of* (46·1) *which is a continuous function of* τ *and* μ *for* $\tau > 0$, $-1 \leqslant \mu \leqslant 1$, *and which is such that*

$$\left.\begin{aligned} J(\tau,\mu) &= O(\tau) \quad \text{as} \quad \tau \to \infty, \\ J(\tau,\mu) &= O(\ln\tau^{-1}) \quad \text{as} \quad \tau \to +0, \end{aligned}\right\} \tag{46·2}$$

uniformly for $-1 \leqslant \mu \leqslant 1$. *Then* $J(\tau,\mu) = 0$.

Under the conditions of the theorem a non-negative function $\phi(\tau)$ of the class $C(\ln \tau^{-1}, \tau)$ exists such that $|J(\tau,\mu)| \leqslant \phi(\tau)$ for $\tau > 0$, $-1 \leqslant \mu \leqslant 1$, and hence, by (46·1),

$$|J(\tau,\mu)| = |\Lambda_{\tau,\mu}\{J(t,\mu')\}| \leqslant \Lambda_{\tau,\mu}\{\phi(t)\}. \qquad (46\cdot3)$$

Since $|P_n(\mu)| \leqslant 1$, therefore

$$p(\mu,\mu') \leqslant \sum_{n=0}^{N} |\omega_n| = W \quad \text{(say)}. \qquad (46\cdot4)$$

Hence, on inverting the order of integration, it is found that

$$\Lambda_{\tau,\mu}\{\phi(t)\} \leqslant \tfrac{1}{2} W \int_0^\infty \phi(t)\, E_1(|t-\tau|)\, dt = W\Lambda_\tau\{\phi(t)\}, \qquad (46\cdot5)$$

where Λ_τ is Hopf's operator. From Theorem 15·1 with $L = \Lambda$ it follows that $\Lambda_\tau\{\phi(t)\} \in C(1,\tau)$. Hence there are positive constants α, β such that

$$\Lambda_\tau\{\phi(t)\} \leqslant \alpha(\tau+\beta), \qquad (46\cdot6)$$

and we can take $\beta > 1$. From (46·3), (46·5) and (46·6),

$$|J(\tau,\mu)| \leqslant W\alpha(\tau+\beta). \qquad (46\cdot7)$$

On evaluating the t-integrals it is found that

$$\begin{aligned}\Lambda_{\tau,\mu}\{t+\beta\} &= \frac{1}{2}\int_0^1 p(\mu,\mu')(\tau+\beta+\mu')\,d\mu' \\ &\quad + \frac{1}{2}\int_0^1 p(\mu,-\mu')[\tau+\beta-\mu'-(\beta-\mu')\exp(-\tau/\mu')]\,d\mu' \\ &\leqslant \frac{1}{2}\int_{-1}^1 p(\mu,\mu')(\tau+\beta+\mu')\,d\mu',\end{aligned}$$

since $(\beta-\mu') > 0$. Hence, by (45·4), (45·5) and (45·6),

$$\Lambda_{\tau,\mu}\{t+\beta\} \leqslant \omega_0(\tau+\beta) + \tfrac{1}{3}\omega_1\mu < \omega_0(\tau+\beta+1) \qquad (46\cdot8)$$

and therefore, by (46·1), (46·7) and (46·8),

$$|J(\tau,\mu)| = |\Lambda_{\tau,\mu}\{J(t,\mu)\}| \leqslant W\alpha\omega_0(\tau+\beta+1). \qquad (46\cdot9)$$

From (46·7) and (46·9) it follows by induction that

$$|J(\tau,\mu)| \leqslant W\alpha\omega_0^n(\tau+\beta+n)$$

for $n = 0, 1, \ldots$. Since $\omega_0^n(\tau+\beta+n) \to 0$ as $n \to \infty$ for each fixed τ, therefore $J(\tau,\mu) = 0$.

46·2. *The conservative case.* We shall first prove

THEOREM 46·2. *Let* $\omega_0 = 1$ *and let* $J(\tau,\mu)$ *be a solution of* (46·1) *which is continuous and non-negative for* $\tau > 0$, $-1 \leqslant \mu \leqslant 1$ *and which is such that*

$$\left.\begin{aligned} J(\tau,\mu) &= O(\ln \tau^{-1}) \quad \text{as} \quad \tau \to +0, \\ J(\tau,\mu) &= O(1) \quad \text{as} \quad \tau \to \infty, \end{aligned}\right\} \tag{46·10}$$

uniformly for $-1 \leqslant \mu \leqslant 1$. *Then* $J(\tau,\mu) = 0$.

As in the preceding proof, $J(\tau,\mu) \leqslant \phi(\tau)$ for $\tau > 0$, $-1 \leqslant \mu \leqslant 1$, where $\phi(\tau) \in C(\ln \tau^{-1}, 1)$. Since $\Lambda_\tau\{\phi(t)\} \in C(1,1)$, it follows from (46·3) and (46·5), which continue to hold, that $J(\tau,\mu)$ is bounded. Let

$$0 \leqslant J(\tau,\mu) \leqslant \alpha \quad (\tau \geqslant 0, -1 \leqslant \mu \leqslant 1). \tag{46·11}$$

Consider the equation (7·8), which is true in the conservative case of the homogeneous equation. The left-hand side ($\mathfrak{J} = J$) is non-negative and, by (46·11), it is not greater than

$$\begin{aligned} \tfrac{1}{2}\alpha \int_0^1 \mu \, d\mu \int_0^\infty \exp(-|t-\tau|/\mu)\, dt &= \tfrac{1}{2}\alpha \int_0^\infty E_3(|t-\tau|)\, dt \\ &= \tfrac{1}{2}\alpha[2E_4(0) - E_4(\tau)] \\ &\leqslant \tfrac{1}{3}\alpha. \end{aligned}$$

But the right-hand side of (7·8) is unbounded as $\tau \to \infty$ if $F \neq 0$. Hence $F = 0$ and therefore (6·10) gives

$$\int_0^1 d\mu \int_0^\infty J(t,\mu) \exp(-t/\mu)\, dt = 0.$$

Since $J(\tau,\mu) \geqslant 0$, therefore $J(\tau,\mu) = 0$.

The following theorem is the analogue of Theorem 17·1:

THEOREM 46·3. *Let* $\omega_0 = 1$ *and let*

$$f_0(\tau,\mu) = \tau + B\mu, \quad g_0(\tau,\mu) = \tau + B\mu + B + 1, \tag{46·12}$$

where

$$B = \omega_1/(3-\omega_1). \tag{46·13}$$

Then the sequences defined for $n \geqslant 1$ *by*

$$f_n(\tau,\mu) = \Lambda^n_{\tau,\mu}\{f_0(t,\mu')\}, \quad g_n(\tau,\mu) = \Lambda^n_{\tau,\mu}\{g_0(t,\mu')\} \tag{46·14}$$

converge, as $n \to \infty$, *to a function* $f(\tau,\mu)$, *which is a solution of* (46·1). *The function* $f(\tau,\mu)$ *is of the form*

$$f(\tau,\mu) = \tau + B\mu + q(\tau,\mu), \quad 0 \leqslant q(\tau,\mu) \leqslant B+1. \tag{46·15}$$

A simple calculation shows that

$$\Lambda_{\tau,\mu}\{t+A+B\mu'\} = \frac{1}{2}\int_0^1 p(\mu,\mu')[\tau+A+(B+1)\mu']\,d\mu'$$
$$+\frac{1}{2}\int_0^1 p(\mu,-\mu')\{\tau+[A-(B+1)\mu'][1-\exp(-\tau/\mu')]\}\,d\mu'. \tag{46·16}$$

On taking $A = 0$ and noting that $B+1>0$, we have

$$\Lambda_{\tau,\mu}\{t+B\mu'\} \geqslant \frac{1}{2}\int_{-1}^1 p(\mu,\mu')[\tau+(B+1)\mu']\,d\mu'$$
$$= \tau+\tfrac{1}{3}\omega_1(B+1)\mu$$

by (45·4) and (45·5). But $\frac{1}{3}\omega_1(B+1) = B$, and hence

$$\Lambda_{\tau,\mu}\{f_0(t,\mu')\} \geqslant f_0(\tau,\mu). \tag{46·17}$$

Again, on taking $A = B+1$ in (46·16), it follows that

$$\Lambda_{\tau,\mu}\{t+B+1+B\mu'\} \leqslant \frac{1}{2}\int_{-1}^1 p(\mu,\mu')[\tau+B+1+(B+1)\mu']\,d\mu'$$
$$= \tau+B+1+\tfrac{1}{3}\omega_1(B+1)\mu$$
$$= \tau+B+1+B\mu.$$

Hence
$$\Lambda_{\tau,\mu}\{g_0(t,\mu')\} \leqslant g_0(\tau,\mu). \tag{46·18}$$

From (46·17), (46·18) and the inequality $f_0<g_0$, we have (cf. Theorem 17·1)

$$f_0 \leqslant f_1 < \ldots < f_n < g_n < \ldots < g_1 \leqslant g_0, \tag{46·19}$$

and, as $n\to\infty$,

$$f_n(\tau,\mu)\to f(\tau,\mu), \quad g_n(\tau,\mu)\to g(\tau,\mu),$$

where
$$f_0(\tau,\mu) < f(\tau,\mu) \leqslant g(\tau,\mu) < g_0(\tau,\mu). \tag{46·20}$$

On letting $n\to\infty$ in

$$f_{n+1}(\tau,\mu) = \Lambda_{\tau,\mu}\{f_n(t,\mu')\},$$

it follows, as in Theorem 17·1, that $f(\tau,\mu)$ and $g(\tau,\mu)$ are solutions of (46·1). If

$$J(\tau,\mu) = g(\tau,\mu) - f(\tau,\mu),$$

then $J(\tau,\mu)$ is continuous, non-negative and bounded, and it is therefore zero by Theorem 46·1. Hence the solution $f(\tau,\mu)$ is unique; and by (46·20) and (46·12) it is of the form given in (46·15).

47. The auxiliary equation

By considering a beam for which the incident intensity is given by (cf. equations (26·1)–(26·4))

$$I_0(\mu') = 2\delta(\mu' - \mu_0) \tag{47·1}$$

where $\delta(\mu)$ is Dirac's delta-function and $0 < \mu_0 \leqslant 1$, the auxiliary equation is found from (5·7) and (5·8) to be†

$$(1-\Lambda)_{\tau,\mu}\{J(t,\mu',\mu_0)\} = p(\mu, -\mu_0)\exp(-\tau/\mu_0), \tag{47·2}$$

where $0 < \mu_0 \leqslant 1$, $-1 \leqslant \mu \leqslant 1$ and $J(\tau, \mu, \mu_0)$ denotes the N-solution, viz.

$$J(\tau,\mu,\mu_0) = \sum_{\nu=0}^{\infty} \Lambda^{\nu}_{\tau,\mu}\{p(\mu', -\mu_0)\exp(-t/\mu_0)\}. \tag{47·3}$$

By (46·4),

$$\Lambda_{\tau,\mu}\{p(\mu', -\mu_0)\exp(-t/\mu_0)\} \leqslant W\Lambda_{\tau,\mu}\{1\}, \tag{47·4}$$

and it is easily shown from (45·7) that

$$\Lambda_{\tau,\mu}\{1\} \leqslant \frac{1}{2}\int_{-1}^{1} p(\mu,\mu')\,d\mu' = \omega_0. \tag{47·5}$$

It follows that

$$\Lambda^{\nu}_{\tau,\mu}\{p(\mu', -\mu_0)\exp(-t/\mu_0)\} \leqslant W\omega_0^{\nu} \quad (\nu = 0, 1, \ldots), \tag{47·6}$$

and the N-series therefore converges if $0 < \omega_0 < 1$. When $\omega_0 = 1$, the question of the convergence has not been completely settled. In § 27 the convergence of $\Sigma L^{\nu}_{\tau}\{\exp(-\sigma t)\}$ for $\psi_0 = \frac{1}{2}$ was a delicate matter depending upon parts of Chapter 4, the extensions of which to the present case are not immediately apparent. Partly for this reason, and also because it is easy to lose the essential structure in the intricacies of rigour, the analysis will from this point be formal, though, as stated above, it can be justified when $0 < \omega_0 < 1$.

47·1. *The 'scattering function'* $S(\mu, \mu_0)$. Let

$$S(\mu,\mu_0) = \int_0^{\infty} J(t,\mu,\mu_0)\exp(-t/\mu)\,dt, \tag{47·7}$$

† In order to avoid a double set of variables, we shall work with μ, μ_0, etc., and not with their reciprocals as in the preceding chapters.

where $0<\mu\leqslant 1$, $0<\mu_0\leqslant 1$. This corresponds to the function $R(s,\sigma)$ of Theorem 28·1 and it satisfies

$$S(\mu,\mu_0)=S(\mu_0,\mu). \tag{47·8}$$

This reciprocity is proved by substituting from (47·3) into (47·7) and integrating term by term. On using the 'symmetrical property' (45·11), which holds when Λ is replaced by Λ^ν, we get (47·8). (Cf. the proof of (28·9).)

47·2. *The integral equation for* $S(\mu,\mu_0)$. From (47·2), by the differentiation formula (45·12) (cf. (28·5)),

$$(1-\Lambda)_{\tau,\mu}\left\{\frac{\partial}{\partial t}J(t,\mu',\mu_0)+\mu_0^{-1}J(t,\mu',\mu_0)\right\}$$
$$=\frac{1}{2}\int_0^1 p(\mu,-u)\,J(0,-u,\mu_0)\exp(-\tau/u)\frac{du}{u}. \tag{47·9}$$

But on multiplying (47·2) (with u in place of μ_0) by

$$\tfrac{1}{2}J(0,-u,\mu_0)/u$$

and integrating with respect to u over $(0,1)$, we get (cf. (28·6) multiplied by $J(0,\sigma)$)

$$(1-\Lambda)_{\tau,\mu}\left\{\frac{1}{2}\int_0^1 J(t,\mu',u)\,J(0,-u,\mu_0)\frac{du}{u}\right\}$$
$$=\frac{1}{2}\int_0^1 p(\mu,-u)\,J(0,-u,\mu_0)\exp(-\tau/u)\frac{du}{u}, \tag{47·10}$$

and hence

$$\frac{\partial}{\partial\tau}J(\tau,\mu,\mu_0)+\mu_0^{-1}J(\tau,\mu,\mu_0)=\frac{1}{2}\int_0^1 J(\tau,\mu,u)\,J(0,-u,\mu_0)\frac{du}{u}, \tag{47·11}$$

assuming that the order of each function is such that there is no non-null solution of the homogeneous equation. Equation (47·11) corresponds to (28·8).

Multiply (47·11) by $\exp(-\tau/\mu)$ $(\mu>0)$ and integrate with respect to τ over $(0,\infty)$, integrating the first term by parts. On using (47·7) and rearranging, we get (cf. (28·10))

$$\left(\frac{1}{\mu}+\frac{1}{\mu_0}\right)S(\mu,\mu_0)=J(0,\mu,\mu_0)+\frac{1}{2}\int_0^1 S(\mu,u)\,J(0,-u,\mu_0)\frac{du}{u}. \tag{47·12}$$

From (47·2) with $\tau = 0$, $0 < \mu_0 \leqslant 1$, $-1 \leqslant \mu \leqslant 1$,

$$J(0, \mu, \mu_0) = \frac{1}{2}\int_0^1 p(\mu, v)\frac{dv}{v}\int_0^\infty J(t, v, \mu_0)\exp(-t/v)\,dt + p(\mu, -\mu_0)$$

$$= \frac{1}{2}\int_0^1 p(\mu, v)\,S(v, \mu_0)\frac{dv}{v} + p(\mu, -\mu_0)$$

$$= \frac{1}{2}\int_0^1 p(\mu, v)\,S(\mu_0, v)\frac{dv}{v} + p(\mu, -\mu_0) \qquad (47·13)$$

by (47·8). This corresponds to (28·11).

It is not possible to find an integral equation for $J(0, \mu, \mu_0)$ from (47·12) and (47·13) (as one was found for $J(0, s)$ in § 28), but on using (45·3) we obtain the following integral equation for $S(\mu, \mu_0)$:

$$\left(\frac{1}{\mu}+\frac{1}{\mu_0}\right)S(\mu, \mu_0) = p(\mu, -\mu_0) + \frac{1}{2}\int_0^1 p(\mu_0, u)\,S(\mu, u)\frac{du}{u} + \frac{1}{2}\int_0^1 p(\mu, v)\,S(\mu_0, v)\frac{dv}{v} + \frac{1}{4}\int_0^1\int_0^1 S(\mu, u)\,p(u, -v)\,S(v, \mu_0)\frac{du\,dv}{uv}, \qquad (47·14)$$

where $0 < \mu \leqslant 1$, $0 < \mu_0 \leqslant 1$. This integral equation is (for axial symmetry) equivalent to that found by Chandrasekhar from principles of invariance. (See [1], chap. IV, equation (28).) The more general equation can be obtained similarly.

48. The reduction to an H-function

On substituting for $p(\mu, v)$ from (45·1) into (47·13), we have

$$J(0, \mu, \mu_0) = \sum_{r=0}^{N} \omega_r P_r(\mu)\,\phi_r(\mu_0) \quad (-1 \leqslant \mu \leqslant 1, 0 < \mu_0 \leqslant 1), \qquad (48·1)$$

where $\phi_r(\mu) = P_r(-\mu) + \frac{1}{2}\int_0^1 P_r(v)\,S(\mu, v)\frac{dv}{v} \quad (0 < \mu \leqslant 1). \qquad (48·2)$

Also, by (45·1) and (48·1), (47·12) can be written

$$\left(\frac{1}{\mu}+\frac{1}{\mu_0}\right)S(\mu, \mu_0) = \sum_{r=0}^{N} \omega_r \phi_r(\mu_0)\left\{P_r(\mu) + \frac{1}{2}\int_0^1 P_r(-u)\,S(\mu, u)\frac{du}{u}\right\}.$$

Since $P_r(-\mu) = (-1)^r P_r(\mu)$, this is (by (48·2))

$$\left(\frac{1}{\mu}+\frac{1}{\mu_0}\right) S(\mu,\mu_0) = \sum_{r=0}^{N} (-1)^r \omega_r \phi_r(\mu)\phi_r(\mu_0). \tag{48·3}$$

On substituting for $S(\mu, v)$ in (48·2), we get

$$\phi_r(\mu) = P_r(-\mu) + \frac{1}{2}\sum_{s=0}^{N}(-1)^s \omega_s \mu\phi_s(\mu)\int_0^1 \frac{P_r(v)\phi_s(v)}{\mu+v}\,dv, \tag{48·4}$$

where $r = 0, 1, \ldots, N$ and $0 < \mu \leqslant 1$. This is a set of $N+1$ integral equations for $\phi_0(\mu), \ldots, \phi_N(\mu)$. We shall show that each function is the product of a polynomial and an H-function.

Since $P_n(u)$ satisfies the recurrence relation

$$(n+1)P_{n+1}(u) - (2n+1)uP_n(u) + nP_{n-1}(u) = 0 \quad (n \geqslant 0), \tag{48·5}$$

therefore, for $n \geqslant 0$,

$$(n+1)P_{n+1}(u) + (2n+1)\mu P_n(u) + nP_{n-1}(u) = (2n+1)(\mu+u)P_n(u). \tag{48·6}$$

Multiply the equations (48·4) in which $r = n-1,\ n,\ n+1$ $(n \leqslant N-1)$ by n, $(2n+1)\mu$, $n+1$, respectively, and add. Since $P_r(-\mu) = (-1)^r P_r(\mu)$, the first terms on the right have a zero sum by (48·5), and on using (48·6) we get

$$(n+1)\phi_{n+1}(\mu) + (2n+1)\mu\phi_n(\mu) + n\phi_{n-1}(\mu) = \tfrac{1}{2}\mu\sum_{s=0}^{N}(-1)^s\omega_s\phi_s(\mu)(2n+1)\int_0^1 P_n(v)\phi_s(v)\,dv. \tag{48·7}$$

This holds for $n = 0, 1, \ldots, N-1$ and $0 < \mu \leqslant 1$. Let

$$p_{rs} = \begin{cases} (-1)^{r-1} + \tfrac{1}{2}\omega_r\displaystyle\int_0^1 P_r(v)\phi_r(v)\,dv & (r = s), \\ \tfrac{1}{2}\omega_s\displaystyle\int_0^1 P_r(v)\phi_s(v)\,dv & (r \neq s). \end{cases} \tag{48·8}$$

Then (48·7) can be written

$$\mu\sum_{s=0}^{N}(-1)^s p_{ns}\phi_s(\mu) = [(n+1)\phi_{n+1}(\mu) + n\phi_{n-1}(\mu)]/(2n+1), \tag{48·9}$$

where $n = 0, 1, \ldots, N-1$ and $0 < \mu \leqslant 1$.

On solving the N homogeneous equations (48·9) for the $N+1$ functions $\phi_r(\mu)$, we get

$$\phi_r(\mu) = q_r(\mu)\,H(\mu) \quad (r = 0, 1, \ldots, N), \tag{48·10}$$

where each $q_r(\mu)$ is a polynomial of degree N and $H(\mu)$ is a function independent of r. The polynomials can be made definite by imposing the condition

$$q_0(0) = 1. \tag{48·11}$$

By (48·9) and (48·10),

$$\mu \sum_{s=0}^{N} (-1)^s p_{ns}\, q_s(\mu) = [(n+1)\,q_{n+1}(\mu) + n q_{n-1}(\mu)]/(2n+1) \tag{48·12}$$

for $n = 0, 1, \ldots, N-1$.

From (48·4), with $r = 0$, and (48·10),

$$q_0(\mu)\,H(\mu) = 1 + \tfrac{1}{2}\mu H(\mu) \sum_{s=0}^{N} (-1)^s \omega_s q_s(\mu) \int_0^1 \frac{q_s(v)\,H(v)}{\mu+v}\,dv \tag{48·13}$$

when $0 < \mu \leqslant 1$. Since

$$[q_s(\mu) - q_s(-v)]/(\mu+v) = \Pi_s(\mu, v), \tag{48·14}$$

where $\Pi_s(\mu, v)$ is a polynomial in μ and v of degree $N-1$, equation (48·13) can be written

$$q_0(\mu)\,H(\mu) = 1 + \mu H(\mu) \int_0^1 \frac{\Psi_{2N}(v)\,H(v)}{\mu+v}\,dv + \tfrac{1}{2}\mu H(\mu)\,Q_{N-1}(\mu), \tag{48·15}$$

where
$$\Psi_{2N}(v) = \frac{1}{2} \sum_{s=0}^{N} (-1)^s \omega_s q_s(v)\, q_s(-v), \tag{48·16}$$

and
$$Q_{N-1}(\mu) = \sum_{s=0}^{N} (-1)^s \omega_s \int_0^1 \Pi_s(\mu, v)\, q_s(v)\, H(v)\, dv. \tag{48·17}$$

$\Psi_{2N}(v)$ is an even polynomial of degree $2N$, and $Q_{N-1}(\mu)$ is a polynomial of degree $N-1$. We shall show that

$$q_0(\mu) = 1 + \tfrac{1}{2}\mu Q_{N-1}(\mu), \tag{48·18}$$

so that (48·15) reduces to the H-equation

$$H(\mu) = 1 + \mu H(\mu) \int_0^1 \frac{\Psi_{2N}(v)\,H(v)}{\mu+v}\,dv. \tag{48·19}$$

From (48·4), with $r = N$, and (48·10), we have

$$\frac{q_N(\mu)}{\mu P_N(-\mu)} = \frac{1}{\mu H(\mu)} + \frac{1}{2}\sum_{s=0}^{N}(-1)^s \omega_s \frac{q_s(\mu)}{P_N(-\mu)} \int_0^1 \frac{P_N(v)\, q_s(v)\, H(v)}{\mu+v}\, dv. \tag{48·20}$$

This equation can be used to define $H(\mu)$ for values of μ in the complex plane cut along $(-1, 0)$ and then the other equations (48·4) will be similarly extended by analytic continuation. Since each $q_s(\mu)$ is a polynomial of degree N, it follows from (48·20) that

$$\lim_{\mu\to\infty} \frac{1}{\mu H(\mu)} = 0. \tag{48·21}$$

From (48·15) we now have

$$\lim_{\mu\to\infty} \mu^{-1}\{q_0(\mu) - \tfrac{1}{2}\mu Q_{N-1}(\mu)\} = 0. \tag{48·22}$$

But $q_0(\mu) - \frac{1}{2}\mu Q_{N-1}(\mu)$ is a polynomial, and it can only be a constant which, by (48·11), is unity. This proves (48·18) and therefore (48·19).

In § 48·2 an expression for $\Psi_{2N}(\mu)$ depending only on $\omega_0, \ldots, \omega_N$ will be found (see (48·42) et seq.). For the theory of Chapter 2 to apply, we want $\Psi_{2N}(\mu) \geqslant 0$ for $0 \leqslant \mu \leqslant 1$, and this condition may introduce restrictions on the coefficients ω_s. When it is satisfied, it will be shown that the condition (8·2) is also satisfied.

48·1. *Relations between the coefficients* p_{rs}. In this subsection and the following one, we shall assume that $\omega_r \neq 0$ $(0 \leqslant r \leqslant N)$. If any $\omega_r = 0$, the appropriate value of $\Psi_{2N}(\mu)$ is easily found from (48·42).

We shall write

$$d_r = 1 - \omega_r/(2r+1). \tag{48·23}$$

By (45·6), $d_r > 0$ for $r \geqslant 1$, $0 \leqslant d_0 < 1$, and $d_0 = 0$ only when $\omega_0 = 1$. With this notation we have the following relations between the coefficients p_{rs}:

For $m, n = 0, 1, \ldots, N$,

$$\sum_{r=0}^{N} \frac{(-1)^r}{\omega_r} p_{mr} p_{nr} = \begin{cases} (-1)^n d_n/\omega_n & (m = n), \\ 0 & (m \neq n), \end{cases} \tag{48·24}$$

and, if $\omega_0 \neq 1$ (so that $d_0 \neq 0$),

$$\sum_{r=0}^{N} (-1)^r \frac{\omega_r}{d_r} p_{rm} p_{rn} = \begin{cases} (-1)^n \omega_n & (m = n), \\ 0 & (m \neq n). \end{cases} \tag{48·25}$$

If $m = 0, 1, \ldots, N-2$ and $n = m+1, m+2, \ldots, N-1$, then

$$\frac{n+1}{2n+1} \frac{p_{m,n+1}}{\omega_{n+1}} + \frac{n}{2n+1} \frac{p_{m,n-1}}{\omega_{n-1}} - \frac{m+1}{2m+1} \frac{p_{n,m+1}}{\omega_{m+1}} - \frac{m}{2m+1} \frac{p_{n,m-1}}{\omega_{m-1}}$$

$$= \begin{cases} c_m & (n = m+1), \\ 0 & (n \neq m+1), \end{cases} \tag{48·26}$$

where $c_m = (-1)^{m-1}(m+1)$

$$\times [\omega_m^{-1}(2m+3)^{-1} + d_{m+1}\, \omega_{m+1}^{-1}(2m+1)^{-1}]. \tag{48·27}$$

In order to prove (48·24), multiply (48·4) (with $r = n$) by $P_m(\mu)$ and integrate over $(0, 1)$. This gives

$$\int_0^1 P_m(\mu)\, \phi_n(\mu)\, d\mu = (-1)^n \int_0^1 P_m(\mu)\, P_n(\mu)\, d\mu$$

$$+ \frac{1}{2} \sum_{s=0}^{N} (-1)^s \omega_s \int_0^1 \int_0^1 \mu P_m(\mu)\, \phi_s(\mu)\, P_n(v)\, \phi_s(v) \frac{d\mu\, dv}{\mu + v}.$$

Interchange μ and v, m and n, and add. Since

$$[(-1)^m + (-1)^n] \int_0^1 P_m(\mu)\, P_n(\mu)\, d\mu$$

$$= \begin{cases} 2(-1)^n/(2n+1) & (m = n), \\ 0 & (m \neq n), \end{cases} \tag{48·28}$$

when $m \neq n$ we get, by (48·8),

$$\frac{2}{\omega_n} p_{mn} + \frac{2}{\omega_m} p_{nm} = \frac{1}{2} \sum_{s=0}^{N} (-1)^s \omega_s \int_0^1 P_m(\mu)\, \phi_s(\mu)\, d\mu \int_0^1 P_n(v)\, \phi_s(v)\, dv$$

$$= 2 \sum_{s=0}^{N} (-1)^s \omega_s^{-1} p_{ms} p_{ns} + 2\omega_m^{-1} p_{nm} + 2\omega_n^{-1} p_{mn},$$

and this is the case $m \neq n$ of (48·24). The case $m = n$ is obtained similarly. The relations (48·25) follow by matrix algebra.†

† From (48·38) (below) we get the equivalent matrix equation

$$\mathbf{P}' \operatorname{diag}(\omega_0/d_0, -\omega_1/d_1, \ldots)\, \mathbf{Q} = \mathbf{I}.$$

To prove (48·26), multiply (48·9) by $P_m(\mu)$ and integrate over (0, 1). On replacing $\mu P_m(\mu)$ on the left by

$$\{(m+1)\,P_{m+1}(\mu)+mP_{m-1}(\mu)\}/(2m+1),$$

we get

$$\sum_{s=0}^{N}(-1)^s p_{ns}\left\{\frac{m+1}{2m+1}\int_0^1 P_{m+1}(\mu)\,\phi_s(\mu)\,d\mu + \frac{m}{2m+1}\int_0^1 P_{m-1}(\mu)\,\phi_s(\mu)\,d\mu\right\}$$

$$=\frac{n+1}{2n+1}\int_0^1 P_m(\mu)\,\phi_{n+1}(\mu)\,d\mu+\frac{n}{2n+1}\int_0^1 P_m(\mu)\,\phi_{n-1}(\mu)\,d\mu.$$

If $m=0,1,\ldots,N-2$ and $n=m+2,m+3,\ldots,N-1$, this reduces to the case $n\neq m+1$ of (48·26), and when $n=m+1$ we get the second part.

48·2. *The characteristic function.* Let the $(N+1)\times(N+1)$ matrices $\mathbf{P},\mathbf{Q},\mathbf{A},\mathbf{B}$ be defined by

$$\mathbf{P}=[p_{rs}],\quad \mathbf{Q}=[(-1)^s p_{rs}/\omega_s], \tag{48·29}$$

$$\mathbf{A}=[a_{rs}],\quad \mathbf{B}=[(-1)^s a_{rs}/\omega_s], \tag{48·30}$$

where
$$\left.\begin{aligned} a_{rs}&=0 \text{ if } s\neq r-1 \text{ or } r+1,\\ a_{r,r-1}&=\frac{(-1)^{r-1}r}{2r+1},\quad a_{r,r+1}=(-1)^{r-1}\frac{(r+1)}{2r+1},\end{aligned}\right\} \tag{48·31}$$

and the subscripts r,s run from 0 to N. Then it follows from the equations (48·12) that

$\omega_s q_s(\mu)=k_1\times$ cofactor of row $N+1$, column $s+1$ in $|\mu\mathbf{Q}-\mathbf{B}|$,
$(-1)^s q_s(\mu)=k_2\times$ cofactor of row $N+1$, column $s+1$ in $|\mu\mathbf{P}-\mathbf{A}|$,
where k_1 and k_2 are constants. Since $q_0(0)=1$,

$$\omega_0=k_1\times \text{cofactor of row } N+1, \text{ column 1 in } |-\mathbf{B}|;$$

hence
$$k_1=\omega_0\,\omega_1\ldots\omega_N(2N)!/2^N(N!)^2. \tag{48·32}$$

Similarly, it is found that

$$k_2=(-1)^{\frac{1}{2}N(N+1)}(2N)!/2^N(N!)^2. \tag{48·33}$$

From these results it follows that the column vector

$$\{\omega_0\,q_0(\mu),\omega_1\,q_1(\mu),\ldots,\omega_N\,q_N(\mu)\}$$
$$=k_1\times \text{column } N+1 \text{ in } (\mu\mathbf{Q}-\mathbf{B})^*, \tag{48·34}$$

where the asterisk denotes the adjugate matrix; and the row vector

$$[q_0(\mu), -q_1(\mu), \ldots, (-1)^N q_N(\mu)]$$
$$= k_2 \times \text{row } N+1 \text{ in } (\mu\mathbf{P}'-\mathbf{A}')^*, \quad (48\cdot35)$$

the prime denoting the transpose. Hence

$$[q_0(-\mu), -q_1(-\mu), \ldots, (-1)^N q_N(-\mu)]$$
$$= (-1)^N k_2 \times \text{row } N+1 \text{ in } (\mu P'+A')^*. \quad (48\cdot36)$$

On pre-multiplying (48·34) by (48·36), we get (from (48·16))

$$\begin{aligned} 2\Psi_{2N}(\mu) &= (-1)^N k_1 k_2 \{(\mu\mathbf{P}'+\mathbf{A}')^* (\mu\mathbf{Q}-\mathbf{B})^*\}_{(N+1),\,(N+1)} \\ &= (-1)^N k_1 k_2 \{(\mu\mathbf{Q}-\mathbf{B})(\mu\mathbf{P}'+\mathbf{A}')\}^*_{(N+1),\,(N+1)} \\ &= (-1)^N k_1 k_2 \times \text{cofactor of element in } (N+1)\text{th row} \\ &\qquad \text{and column of } |\mu^2\mathbf{QP}'+\mu(\mathbf{QA}'-\mathbf{BP}')-\mathbf{BA}'|. \end{aligned} \quad (48\cdot37)$$

On using the equations (48·24) it is found that

$$\mathbf{QP}' = \operatorname{diag}\left(\frac{d_0}{\omega_0}, -\frac{d_1}{\omega_1}, \ldots, (-1)^N \frac{d_N}{\omega_N}\right), \quad (48\cdot38)$$

and the equations (48·26) give

$$\mathbf{QA}'-\mathbf{BP}' = \begin{bmatrix} 0 & c_0 & 0 & 0 & \ldots \\ -c_0 & 0 & c_1 & 0 & \ldots \\ 0 & -c_1 & 0 & c_2 & \ldots \\ \ldots & \ldots & \ldots & \ldots & \ldots \\ \ldots & \ldots & \ldots & \ldots & \ldots \end{bmatrix}, \quad (48\cdot39)$$

where c_m is defined by (48·27) for $m \leqslant N-2$ and

$$c_{N-1} = -\frac{N}{2N-1}\frac{p_{NN}}{\omega_N} - \frac{(N-1)}{2N-1}\frac{p_{N,\,N-2}}{\omega_{N-2}} + \frac{N}{2N+1}\frac{p_{N-1,\,N-1}}{\omega_{N-1}}.$$

Also
$$\mathbf{BA}' = \left[\sum_{i=0}^{N} (-1)^i \omega_i^{-1} a_{ri} a_{si}\right] = [\alpha_{rs}], \quad (48\cdot40)$$

and it is easily seen from (48·31) that $\alpha_{rs} = 0$ unless $s = r-2, r$, or $r+2$, and that

$$\left.\begin{aligned} \alpha_{rr} &= \frac{(-1)^{r-1}}{(2r+1)^2}\left\{\frac{r^2}{\omega_{r-1}} + \frac{(r+1)^2}{\omega_{r+1}}\right\}, \\ \alpha_{r,\,r+2} = \alpha_{r+2,\,r} &= \frac{(-1)^{r-1}}{\omega_{r+1}} \cdot \frac{(r+1)(r+2)}{(2r+1)(2r+5)}. \end{aligned}\right\} \tag{48·41}$$

From (48·32), (48·33) and (48·37)–(48·41) we obtain (after changing the signs of the even rows)

$$\Psi_{2N}(\mu) = \frac{[(2N)!]^2}{2^{2N+1}(N!)^4}\,\omega_0 \ldots \omega_N\,\Delta, \tag{48·42}$$

where Δ is the Nth order determinant

$$\Delta = \begin{vmatrix} \mu^2\omega_0^{-1}d_0 - \alpha_{00} & c_0\mu & -\alpha_{02} & 0 & \cdots \\ c_0\mu & \mu^2\omega_1^{-1}d_1 + \alpha_{11} & -c_1\mu & \alpha_{13} & \cdots \\ -\alpha_{02} & -c_1\mu & \mu^2\omega_2^{-1}d_2 - \alpha_{22} & c_2\mu & \cdots \\ 0 & \alpha_{13} & c_2\mu & \mu^2\omega_3^{-1}d_3 + \alpha_{33} & \cdots \\ \cdots & \cdots & \cdots & \cdots & \cdots \end{vmatrix}. \tag{48·43}$$

Since c_{N-1} does not appear in Δ, the numbers c_m and α_{rs} are given by (48·27) and (48·41). Thus the coefficients in $\Psi_{2N}(\mu)$ are determined by $\omega_0, \ldots, \omega_N$.

It remains to prove that, when $\Psi_{2N}(\mu) \geqslant 0$, it satisfies the condition (8·2). We shall show that†

$$\psi_0 \equiv \int_0^1 \Psi_{2N}(\mu)\,d\mu = \tfrac{1}{2}\{1 - d_0 d_1 \ldots d_N\}, \tag{48·44}$$

and the condition (8·2) then follows since every $d_r \geqslant 0$. Subject to the condition (45·6), the conservative case of the H-equation can only occur when $\omega_0 = 1$.

A direct verification of (48·44) is wellnigh impossible, though it can be done for $N = 1, 2$. We therefore proceed as follows: In Theorem 12, working from the H-equation and without using (8·2), it is proved that (see (12·9) and (12·8))

$$\lim_{\mu\to\infty} 1/H(\mu) = 1 - h_0 = \pm(1 - 2\psi_0)^{\frac{1}{2}}, \tag{48·45}$$

† Kuščer [1].

and this is therefore true for the H-function of this section. Let

$$\lim_{\mu\to\infty} \mu^{-N} q_s(\mu) = l_s \quad (s = 0, 1, \ldots, N). \tag{48·46}$$

By (48·35),

$$\begin{aligned}[l_0, -l_1, \ldots, (-1)^N l_N] &= k_2 \times \text{row } (N+1) \text{ in } (\mathbf{P}')^* \\ &= k_2[P_{N0}, P_{N1}, \ldots, P_{NN}],\end{aligned} \tag{48·47}$$

where P_{rs} is the cofactor of p_{rs} in $|\mathbf{P}|$. By (48·4) (with $r = N$) and (48·10),

$$\begin{aligned}\mu^{-N} q_N(\mu) &= \mu^{-N} \frac{P_N(-\mu)}{H(\mu)} \\ &\quad + \frac{1}{2}\sum_{s=0}^{N} (-1)^s \omega_s \mu^{-N} q_s(\mu) \int_0^1 \frac{P_N(v)\phi_s(v)}{1+v/\mu}\, dv,\end{aligned}$$

and on letting $\mu \to \infty$ we get

$$\begin{aligned}l_N &= \pm(1-2\psi_0)^{\frac{1}{2}} (-1)^N \frac{(2N)!}{2^N (N!)^2} \\ &\quad + \frac{1}{2}\sum_{s=0}^{N} (-1)^s \omega_s l_s \int_0^1 P_N(v)\,\phi_s(v)\, dv,\end{aligned}$$

i.e. by (48·8)

$$\sum_{s=0}^{N} (-1)^s p_{Ns} l_s = \pm(1-2\psi_0)^{\frac{1}{2}} (2N)!/2^N (N!)^2. \tag{48·48}$$

On substituting for $(-1)^s l_s$ from (48·47) and using the value of k_2 given by (48·33), we get

$$(1-2\psi_0)^{\frac{1}{2}} = \pm|\mathbf{P}|. \tag{48·49}$$

From (48·29) it is seen that

$$|\mathbf{Q}| = (-1)^{\frac{1}{2}N(N+1)} |\mathbf{P}|/\omega_0 \ldots \omega_N,$$

and hence (48·38) gives

$$|\mathbf{P}|^2 = d_0 d_1 \ldots d_N. \tag{48·50}$$

Equation (48·44) now follows from (48·49) and (48·50).

48·3. *The scattering function.* From (48·3) and (48·10) we have

$$(\mu^{-1}+\mu_0^{-1})\, S(\mu,\mu_0) = H(\mu)\, H(\mu_0) \sum_{s=0}^{N} (-1)^s \omega_s q_s(\mu)\, q_s(\mu_0). \tag{48·51}$$

The sum on the right is similar in form to that given in (48·16) for $\Psi_{2N}(\mu)$, and it can be evaluated by the method of § 48·2. Corresponding to (48·37), it is found that

$$(\mu^{-1}+\mu_0^{-1})\,S(\mu,\mu_0) = k_1 k_2 H(\mu) H(\mu_0) \times \text{cofactor of element in } (N+1)\text{th row and column of } |\mu\mu_0\mathbf{QP}'-\mu\mathbf{QA}'-\mu_0\mathbf{BP}'+\mathbf{BA}'|. \tag{48·52}$$

The matrices $\mathbf{QP}'$ and $\mathbf{BA}'$ are given by (48·38) and (48·40), but $\mathbf{QA}'$ and $\mathbf{BP}'$ cannot be obtained in a form depending only on $\omega_0, \ldots, \omega_N$. However, on writing (48·8) in the form

$$p_{rs} = \begin{cases} (-1)^{r-1}+\frac{1}{2}\omega_r\int_0^1 P_r(v)\,q_r(v)\,H(v)\,dv & (r=s), \\ \frac{1}{2}\omega_s\int_0^1 P_r(v)\,q_s(v)\,H(v)\,dv & (r\neq s), \end{cases} \tag{48·53}$$

and substituting the polynomial expressions for $P_r(v)$ and $q_s(v)$, the coefficients p_{rs} can be determined in terms of the moments $\alpha_0, \alpha_1, \ldots$ of $H(\mu)$. Thus the scattering function can be found.

49. The law of diffuse reflexion

In an atmosphere subject to known incident radiation, the intensity of the radiation emerging from the surface can often be found, when the scattering function is known, by what Chandrasekhar has called 'the law of diffuse reflexion'.

Let radiation of intensity $I_0(\mu')$ $(0<\mu'\leqslant 1)$ fall on the surface of a semi-infinite atmosphere in which the emission is given by $B_1(\tau)$ and let the phase function be $p(\mu,\mu')$. It follows from (5·7) and (5·8) that the integral equation for the source function $\mathfrak{J}(\tau,\mu)$ is

$$(1-\Lambda)_{\tau,\mu}\{\mathfrak{J}(t,\mu')\} = B_1(\tau)+\frac{1}{2}\int_0^1 p(\mu,-\mu')\,I_0(\mu')\exp(-\tau/\mu')\,d\mu'. \tag{49·1}$$

Let $\mathfrak{J}_1(\tau,\mu)$ be the source function for the same atmosphere when there is no incident radiation, so that

$$(1-\Lambda)_{\tau,\mu}\{\mathfrak{J}_1(t,\mu')\} = B_1(\tau). \tag{49·2}$$

On multiplying the auxiliary equation (47·2) by $\frac{1}{2}I_0(\mu_0)$ and integrating with respect to μ_0 over (0, 1), we get

$$(1-\Lambda)_{\tau,\mu}\left\{\frac{1}{2}\int_0^1 I_0(\mu_0)\,J(t,\mu',\mu_0)\,d\mu_0\right\}$$
$$=\frac{1}{2}\int_0^1 p(\mu,-\mu_0)\,I_0(\mu_0)\exp(-\tau/\mu_0)\,d\mu_0. \qquad (49\cdot3)$$

From (49·1)–(49·3) it is seen that

$$\mathfrak{J}(\tau,\mu)=\mathfrak{J}_1(\tau,\mu)+\frac{1}{2}\int_0^1 I_0(\mu')\,J(\tau,\mu,\mu')\,d\mu' \qquad (49\cdot4)$$

(for $\mathfrak{J}=\mathfrak{J}_1$ when $I_0(\mu')=0$, and there can therefore be no added solution of the homogeneous equation).

Let

$$I(0,+\mu)=\mathfrak{L}_{1/\mu}\{\mathfrak{J}(t,\mu)\}, \qquad (49\cdot5)$$

$$I_1(0,+\mu)=\mathfrak{L}_{1/\mu}\{\mathfrak{J}_1(t,\mu)\}, \qquad (49\cdot6)$$

so that $I(0,+\mu)$ is the required emergent intensity, and $I_1(0,+\mu)$ is that when there is no incident radiation. Then operating on (49·4) by $\mathfrak{L}_{1/\mu}$ and using (47·7), we get

$$I(0,+\mu)=I_1(0,+\mu)+\frac{1}{2\mu}\int_0^1 I_0(\mu')\,S(\mu,\mu')\,d\mu'. \qquad (49\cdot7)$$

This is the law of diffuse reflexion.

49·1. *An atmosphere with constant net flux.* The law of diffuse reflexion has been used by Chandrasekhar to obtain an expression for the emergent intensity in a non-emitting atmosphere with constant net flux F and no incident radiation. Such an atmosphere is conservative (see § 6).

By Theorem 46·3, the source function $\mathfrak{J}_1(\tau,\mu)$ for the atmosphere is given by

$$\mathfrak{J}_1(\tau,\mu)=A[\tau+B\mu+q(\tau,\mu)], \qquad (49\cdot8)$$

where $B=\omega_1/(3-\omega_1)$, $0\leqslant q(\tau,\mu)\leqslant B+1$ and A is a constant depending upon F. The equation (7·8) holds (with $\mathfrak{J}=\mathfrak{J}_1$), and on dividing by τ we get

$$\tfrac{1}{4}F(1-\tfrac{1}{3}\omega_1)=\lim_{\tau\to\infty}\tfrac{1}{2}\tau^{-1}\int_0^1\mu\,d\mu\int_0^\infty At\exp(-|t-\tau|/\mu)\,dt.$$

The terms omitted, being bounded, give a zero limit. On evaluating the integral we get

$$\tfrac{1}{4}F(1-\tfrac{1}{3}\omega_1) = \lim_{\tau\to\infty} \tfrac{1}{2}A\tau^{-1}[\tfrac{2}{3}\tau + E_5(\tau)] = \tfrac{1}{3}A,$$

and hence

$$\mathfrak{J}_1(\tau,\mu) = \tfrac{3}{4}F[(1-\tfrac{1}{3}\omega_1)\tau + \tfrac{1}{3}\omega_1\mu + (1-\tfrac{1}{3}\omega_1)q(\tau,\mu)]. \qquad (49\cdot9)$$

Now consider the atmosphere for which the source function is

$$\mathfrak{J}(\tau,\mu) = \tfrac{3}{4}F[(1-\tfrac{1}{3}\omega_1)\tau + \tfrac{1}{3}\omega_1\mu]. \qquad (49\cdot10)$$

Since $\mathfrak{J}(\tau,\mu)$ and $\mathfrak{J}_1(\tau,\mu)$ differ only by a non-negative bounded function, and since (by Theorem 46·2) there is no non-negative bounded solution of the homogeneous Milne equation other than zero, $\mathfrak{J}(\tau,\mu)$ must be a source function for the same non-emitting atmosphere subject to certain incident radiation $I_0(\mu')$ $(0<\mu'\leqslant 1)$. From (46·16), on using (45·4) and (45·5), it is easily found that

$$(1-\Lambda)_{\tau,\mu}\{\mathfrak{J}(t,\mu')\} = -\tfrac{3}{8}F\int_0^1 p(\mu,-\mu')\,\mu'\exp(-\tau/\mu')\,d\mu', \qquad (49\cdot11)$$

and on comparing this with (49·1) we get

$$I_0(\mu') = -\tfrac{3}{4}F\mu', \qquad (49\cdot12)$$

a result which is possible mathematically if not physically. From (49·5) and (49·10),

$$I(0,+\mu) = \tfrac{3}{4}F[(1-\tfrac{1}{3}\omega_1)\mu + \tfrac{1}{3}\omega_1\mu] = \tfrac{3}{4}F\mu, \qquad (49\cdot13)$$

and hence (49·7) gives

$$I_1(0,+\mu) = \tfrac{3}{4}F\left\{\mu + \frac{1}{2\mu}\int_0^1 \mu'\,S(\mu,\mu')\,d\mu'\right\}. \qquad (49\cdot14)$$

This will give the emergent intensity in the given atmosphere when the scattering function is known.

APPENDIX

NOTES ON FURTHER PROBLEMS

50. Polarization

When polarization of the radiation field is assumed, the phase function is replaced by a phase matrix, the intensity by a matrix function, and the equation of transfer by a matrix equation. Details are given in Chandrasekhar [1], where a solution is found by the method of discrete ordinates. The analysis involved in such a solution is always long and the convergence of the nth approximation to the solution as $n \to \infty$ has, so far, only been proved in the simplest case, viz. that of isotropic scattering. (See Anselone [1].)

From the matrix equation of transfer, the matrix equivalent of Milne's equation can easily be found, and a rigorous solution using the Ambartsumian technique should therefore be possible.

51. Inhomogeneous atmospheres

When the phase function is $\omega_0(\tau)$ (see (2·10)), the Milne equations for finite and semi-infinite atmospheres with axial symmetry are obtained on replacing ω_0 by $\omega_0(\tau)$ in (4·8) and (4·10), (5·11) and (5·12).

A formal solution in the case of the semi-infinite atmosphere was first obtained by Sobolev [3], and that for a finite atmosphere by Ueno [4]. By letting $\tau_1 \to \infty$, Ueno has deduced Sobolev's solution from his own.

Ueno makes use of two auxiliary equations

$$p(\mu, \tau, \tau_1) = \omega_0(\tau)\,\bar{\Lambda}_\tau\{p(\mu, t, \tau_1)\} + \omega_0(\tau)\exp(-\tau/\mu), \tag{51·1}$$

$$p^*(\mu, \tau, \tau_1) = \omega_0(\tau_1 - \tau)\,\bar{\Lambda}_\tau\{p^*(\mu, t, \tau_1)\} + \omega_0(\tau_1 - \tau)\exp(-\tau/\mu), \tag{51·2}$$

where $\bar{\Lambda}$ is defined by (4·9). The function $p(\mu, \tau, \tau_1)$ is the source function due to a pencil of radiation incident at an angle $\cos^{-1}\mu$ on the surface $\tau = 0$; $p^*(\mu, \tau_1 - \tau, \tau_1)$ is that due to a similar pencil incident on the surface $\tau = \tau_1$. In order to avoid derivatives of $\omega_0(\tau)$, the auxiliary equations are differentiated with respect to τ_1 (cf. § 40·2) and the solution is found to depend on four functions $X(\mu,\tau_1)$, $Y(\mu,\tau_1)$, $X^*(\mu,\tau_1)$, $Y^*(\mu,\tau_1)$, which satisfy the relations

$$Y(\mu,\tau_1) = \exp(-\tau_1/\mu)\, X(-\mu,\tau_1), \tag{51·3}$$

$$Y^*(\mu,\tau_1) = \exp(-\tau_1/\mu)\, X^*(-\mu,\tau_1), \tag{51·4}$$

and the integral equations

$$X^*(\mu,\tau_1) = 1 + \frac{1}{2}\int_0^1 \frac{d\mu'}{\mu'} \int_0^{\tau_1} \omega_0(\tau)\, X^*(\mu,\tau)\, X^*(\mu',\tau) \times \exp[-(\tau_1-\tau)(1/\mu+1/\mu')]\, d\tau, \tag{51·5}$$

$$X(\mu,\tau_1) = 1 + \frac{1}{2}\int_0^1 \frac{d\mu'}{\mu'} \int_0^{\tau_1} \omega_0(\tau)\, Y^*(\mu,\tau)\, Y^*(\mu',\tau)\, d\tau. \tag{51·6}$$

The analysis is easily justified by the methods of Chapters 7 and 8 if X, Y, X^*, Y^* are defined by

$$\omega_0(0)\, X(\mu,\tau_1) = p(\mu,0,\tau_1), \quad \omega_0(0)\, Y(\mu,\tau_1) = p^*(\mu,\tau_1,\tau_1), \tag{51·7}$$

$$\omega_0(\tau_1)\, X^*(\mu,\tau_1) = p^*(\mu,0,\tau_1), \quad \omega_0(\tau_1)\, Y^*(\mu,\tau_1) = p(\mu,\tau_1,\tau_1), \tag{51·8}$$

where $p(\mu,\tau,\tau_1)$ and $p^*(\mu,\tau_1,\tau_1)$ are the N-solutions of (51·1) and (51·2), but in proceeding to the limit as $\tau_1 \to \infty$ (and also in Sobolev's solution) assumptions have to be made whose physical implications are obscure.

52. The formation of absorption lines

A theory of the formation of absorption lines based on radiative-transfer theory was first worked out in detail by Eddington [2]. In this classical paper he said: 'The crucial question is whether light absorbed in one part of the line is re-emitted in the same part of the line. If so, the blackening in this frequency is independent of what is happening in neighbouring frequencies. The alternative is that the re-emission has a probability distribution and is correlated to, but not determined by, the absorbed frequency. . . . In that case the line can only be studied as a whole.' That there must be a redistribution in frequency is now generally accepted, and since about 1949, several papers have dealt with partially or completely 'non-coherent scattering'. (See, for example, Savedoff [1], Sobolev [2], Busbridge [1] and [4].) Unfortunately, the complications of the analysis make numerical computations of exact solutions difficult. Recent work has been concerned with the introduction of simplifications and approximations. (See, for example, Miyamoto [1].)

The literature on the subject of absorption lines is very large, and the theory has several variants. Most of these are given by Unsöld in [1], who also lists most of the papers published before 1953.

53. Time-dependent problems

In astrophysics, radiation fields are usually stationary, but the diffusion of neutrons in a medium is likely to vary with the time t. (See Davison [1].)

The simplest form of the equation of transfer for a time-dependent problem is

$$\lambda \frac{\partial}{\partial t} I(\tau, \mu, t) + \mu \frac{\partial}{\partial \tau} I(\tau, \mu, t) = I(\tau, \mu, t) - \tfrac{1}{2}\omega_0 \int_{-1}^{1} I(\tau, \mu', t)\, d\mu', \tag{53·1}$$

where λ is usually a small constant. Let

$$\bar{I}(\tau, \mu, s) = \mathfrak{L}_s\{I(\tau, \mu, t)\} = s \int_0^\infty I(\tau, \mu, t) \exp(-st)\, dt, \tag{53·2}$$

where $\operatorname{re} s > \alpha$ (say). On operating on (53·1) by $\mathfrak{L}_s$, it is found that

$$\mu \frac{\partial}{\partial \tau} \bar{I}(\tau, \mu, s) = (1 - \lambda s) \bar{I}(\tau, \mu, s) + \lambda s I(\tau, \mu, 0) - \tfrac{1}{2}\omega_0 \int_{-1}^{1} \bar{I}(\tau, \mu', s)\, d\mu'. \tag{53·3}$$

This is similar in form to the equation of transfer in a stationary field. When (53·3) has been solved for $\bar{I}(0, \mu, s)$, the emergent intensity will be given by

$$I(0, \mu, t) = \mathfrak{L}_t^{-1}\{\bar{I}(0, \mu, s)\}. \tag{53·4}$$

The difficulties of this method are considerable and a direct attack on (53·1) is to be preferred. This has been done by J. Lehner and G. M. Wing in [1] and [2]. Their solution employs advanced-function theory and the theory of semigroups of operators. A simpler version is needed for practical work.

54. Spherical atmospheres

For most purposes in astrophysics, the solution of the equation of transfer for the plane parallel atmosphere gives a good approximation to the actual reality, but this breaks down near to the limit of the sun. It is also unsatisfactory in the case of stars with very extended atmospheres (such as supergiants) and planetary nebulae, where the curvature cannot be completely neglected. Recently, D. Barbier and G. W. Curtis [1] have discussed the conditions under which the neglect of curvature can have an appreciable effect.

In neutron-diffusion theory, a spherical medium will usually be homogeneous. The exact solution for an isotropic point source in such a medium has been obtained by Ambartsumian [3] and Davison [1], and Davison has also shown that diffusion in a homogeneous sphere is, under certain conditions, mathematically equivalent to diffusion in an infinite plane slab of thickness equal to twice the radius.

In spherical atmospheres, on the other hand, the density ρ and the absorption coefficient κ will vary with the distance r from the centre, the usual assumption being $\kappa\rho \propto r^{-n}$ $(n > 1)$. Approximate solutions in this case have been obtained by Chandrasekhar [1], using the method of discrete ordinates; by Sen [1], using the spherical harmonic method, which had already been applied in the case of diffusion through a sphere by Marshak [1]; and an iterative method for correcting for curvature has been devised by Proisy [1]. Meanwhile the exact solution of the equation of transfer remains unknown.

REFERENCES

M.N. = *Monthly Notices of the Roy. Astr. Soc.*
Ap. J. = *Astrophysical Journal.*

AMBARTSUMIAN, V. A. [1]. Diffusion of light by planetary atmospheres. *Astr. J. USSR*, **19**, 30, 1942.

[2]. Diffuse reflection of light by a foggy medium. *Dokl. Akad. Nauk, SSSR*, **38**, 229, 1943.

[3]. A point source of light within a scattering medium. *Bull. Erevan Astron. Obs.* no. 6, 1945.

ANSELONE, P. M. [1]. Convergence of the Wick–Chandrasekhar approximation technique in radiative transfer. *Ap. J.* **128**, 124, 1958.

BARBIER, D., and CURTIS, G. W. [1]. Remarques sur le problème sphérique de transfert dans les atmosphères stellaires. *Ann. Astrophys.* **19**, 190, 1956.

BUSBRIDGE, I. W. [1]. Coherent and non-coherent scattering in the theory of line formation. *M.N.* **113**, 52, 1953.

[2]. On solutions of the non-homogeneous form of Milne's first integral equation. *Quart. J. Math.* (Oxford, 2), **6**, 218, 1955.

[3]. Finite atmospheres with isotropic scattering. I, *M.N.* **115**, 521, 1955; II, *M.N.* **116**, 304, 1956; III, *M.N.* **117**, 516, 1957.

[4]. A mathematical verification of the principles of invariance as applied to completely non-coherent scattering and to interlocked multiplet lines. *M.N.* **115**, 661, 1955.

[5]. On the *H*-functions of S. Chandrasekhar. *Quart. J. Math.* (Oxford, 2), **8**, 133, 1957.

BUSBRIDGE, I. W. and STIBBS, D. W. N. [1]. On the intensities of interlocked multiplet lines in the Milne–Eddington Model. *M.N.* **114**, 2, 1954.

CARSLAW, H. S. and JAEGER, J. C. [1]. *Operational Methods in Applied Mathematics.* Oxford, 1941.

CHANDRASEKHAR, S. [1]. *Radiative Transfer.* Oxford, 1950.

[2]. On the radiative equilibrium of a stellar atmosphere. XIV. *Ap. J.* **105**, 164, 1947.

[3]. On the radiative equilibrium of a stellar atmosphere. XVII. *Ap. J.* **105**, 441, 1947.

[4]. On the radiative equilibrium of a stellar atmosphere. XXI. *Ap. J.* **106**, 152, 1947; XXII, *Ap. J.* **107**, 48, 1948.

[5]. The transfer of radiation in stellar atmospheres. *Bull. Amer. Math. Soc.* **53**, 641, 1947.

CHANDRASEKHAR, S., ELBERT, D., and FRANKLIN, A. [1]. The *X*- and *Y*-functions for isotropic scattering. I and II. *Ap. J.* **115**, 244, and 269, 1952.

CRUM, M. M. [1]. On an integral equation of Chandrasekhar. *Quart. J. Math.* (Oxford, 1), **18**, 244, 1947.

DAVISON, B. (with SYKES, J. B.) [1]. *Neutron Transport Theory.* Oxford, 1957.

DINGLE, R. B. [1]. The anomalous skin effect and the reflectivity of metals. *Physica,* **19**, 311, 1953.

EDDINGTON, A. S. [1]. *The Internal Constitution of the Stars.* Cambridge, 1930.

[2]. The formation of absorption lines. *M.N.* **89**, 620, 1929.

ELLIOTT, J. P. [1]. Milne's problem with a point source. *Proc. Roy. Soc.* A, **228**, 424, 1955.

HOPF, E. [1]. *Mathematical Problems of Radiative Equilibrium.* Cambridge Tracts, no. 31, 1934.

HORAK, H. G. and LUNDQUIST, C. A. [1]. The transfer of radiation by an emitting atmosphere. I, *Ap. J.* **116**, 477, 1952; II, *Ap. J.* **119**, 42, 1954.

HUANG, S. S. [1]. Some formulae for the emergent intensities by the Laplace transformation. *Ann. Astrophys.* **15**, 352, 1952.

KOURGANOFF, V. (with BUSBRIDGE, I. W.) [1]. *Basic Methods in Transfer Problems.* Oxford, 1952.

KUŠČER, I. [1]. Milne's problem for anisotropic scattering. *J. Math. Phys.* **34**, 256, 1955.

LEHNER, J. and WING, G. M. [1]. On the spectrum of an unsymmetric operator arising in the transport theory of neutrons. *Commun. Pure Appl. Math.* **8**, 217, 1955.

[2]. Solution of the linearized Boltzmann equation for slab geometry. *Duke Math. J.* **23**, 125, 1956.

MARK, C. [1]. The neutron density near a plane surface. *Phys. Rev.* **72**, 558, 1947.

MARSHAK, R. E. [1]. Note on the spherical harmonic method as applied to the Milne problem for a sphere. *Phys. Rev.* **71**, 443, 1947.

MILNE, E. A. [1]. *Handbuch der Astrophysik* (iii), chap. II, 'Thermodynamics of the stars'. J. Springer, Berlin, 1930.

[2]. Radiative equilibrium in the outer layers of a star: the temperature distribution and the law of darkening. *M.N.* **81**, 361, 1921.

MIYAMOTO, S. [1]. On the calculation of non-coherent contours. *Contr. Inst. Astrophys. Kyoto*, no. 46, 1954; no. 57, 1955.

PLACZEK, G. and SEIDEL, W. [1]. Milne's problem in transport theory. *Phys. Rev.* **72**, 550, 1947.

PREISENDORFER, R. W. [1]. A mathematical foundation for radiative transfer theory. *J. Math. Mech.* **6**, 685, 1957.

PROISY, P. [1]. L'effet de la courbure sur l'interprétation de l'assombrissement du disque solaire. *Ann. Astrophys.* **21**, 151, 1958.

REUTER, G. E. H. and SONDHEIMER, E. H. [1]. The theory of the anomalous skin effect in metals. *Proc. Roy. Soc.* A, **195**, 336, 1948.

SAVEDOFF, M. P. [1]. Formation of absorption lines by non-coherent scattering. *Ap. J.* **115**, 509, 1952.

SCHUSTER, A. [1]. Radiation through a foggy atmosphere. *Ap. J.* **21**, 1, 1905.

SCHWARZSCHILD, K. [1]. Über das Gleichgewicht der Sonnenatmosphäre, *Göttinger Nachr.* 41, 1906.

[2]. Über Diffusion und Absorption in der Sonnenatmosphäre. *S.B. preuss. Akad. Wiss.* 1183, 1914.

SEN, H. K. [1]. The radiative equilibrium of a spherical planetary nebula. *Ap. J.* **110**, 276, 1949.

SMITHIES, F. [1]. Singular integral equations. *Proc. Lond. Math. Soc.* (2), **46**, 409, 1940.

SOBOLEV, V. V. [1]. A new method in the theory of the diffusion of light. *Russ. Astr. J.* **28**, 355, 1951.

[2]. The formation of absorption lines with non-coherent scattering of light. *Russ. Astr. J.* **31**, 231, 1954.

[3]. Transfer of radiation in an inhomogeneous medium. *Dokl. Akad. Nauk, SSSR*, **111**, 1000, 1956.

[4]. Diffusion of radiation in a medium of finite optical thickness. *Astr. J. USSR*, **34**, 336, 1957.

[5]. *Transfer of Radiation Energy in the Atmospheres of Stars and Planets.* Moscow, 1956.

SONDHEIMER, E. H. [1]. The theory of the anomalous skin effect in anisotropic metals. *Proc. Roy. Soc.* A, **224**, 260, 1954.

STIBBS, D. W. N. and WEIR, R. E. [1]. On the H-functions for isotropic scattering. *M.N.* **119**, 512, 1959.

TITCHMARSH, E. C. [1]. *The Theory of Functions.* Oxford, 1932.

[2]. *Introduction to the Theory of Fourier Integrals.* Oxford, 1937.

UENO, S. [1]. The formation of absorption lines by coherent and non-coherent scattering. IV. The solution of the equation of transfer by the probabilistic method. *Contr. Inst. Astrophys. Kyoto*, no. 64, 1956.

[2]. The probabilistic method for problems of radiative transfer. III. Line formation by coherent scattering. *J. Math. Mech.* **7**, 629, 1958.

[3]. The probabilistic method for problems of radiative transfer. IX. Diffuse reflection and transmission in a finite atmosphere with isotropic scattering in the non-conservative case. *Ann. Astrophys.* **22**, 468, 1959; XI. On the scattering and transmission functions of S. Chandrasekhar. *Ann. Astrophys.* **22**, 484, 1959.

[4]. Diffuse reflections and transmission in a finite inhomogeneous atmosphere. *Unpublished.*

UNSÖLD, A. [1]. *Physik der Sternatmosphären.* Springer Verlag, Berlin, 1955.

WIDDER, D. V. [1]. *The Laplace Transform.* Princeton, 1946.

WIENER, N. and HOPF, E. [1]. Über eine Klasse singulärer Integralgleichungen. *S.B. preuss. Akad. Wiss.* 696, 1931.

GENERAL INDEX

INDEX OF NOTATIONS